CUTTING YOUR CAR USE

Anna Semlyen's interest in traffic reduction stems from a bicycle crash in 1992 when a car driver failed to stop at a give way sign. Car-free by choice, she has localised her main activities to be close to where she lives. She rides a Brompton folding bicycle with basket or trailer, shares lifts, takes taxis and sometimes hires a car.

Anna has a MA in Philosophy, Politics and Economics from Oxford University and MSc in Health Economics from York. As well as this book, her Cutting Your Car Use consultancy services include integrated travel guides tailored to a location, traffic reduction training, car cost sheets and Green Travel Plans. She is also a Research Associate with Passenger Transport Networks, a consultancy specialising in the use of Geographic Information Systems (e.g. postcode-based mapping). She has written for *Bycycle* magazine.

Anna was born in 1969, is married, and enjoys teaching yoga, juggling and growing organic food. This is her second book.

CUTTING YOUR YOUR CAR USE

SAVE MONEY, BE HEALTHY, BE GREEN!

ANNA SEMLYEN

GREEN BOOKS

First published in June 2000
by Green Books Ltd
Foxhole, Dartington Totnes, Devon TQ9 6EB
greenbooks@gn.apc.org www.greenbooks.co.uk

© Anna Semlyen 2000
info@cuttingyourcaruse.co.uk
www.cuttingyourcaruse.co.uk

Cartoons by Andy Singer © Andrew B. Singer 1992-1999
www.andysinger.com

Cover design by Rick Lawrence
Cover illustration based on a cartoon by Andy Singer

Text printed by J.W. Arrowsmith Ltd, Bristol
on Cyclus Offset 100% recycled paper

British Library Cataloguing in Publication Data
available on request

ISBN 1 870098 87 0

Companies, local authorities or other organisations
considering car reduction programmes who would
like to discuss bulk purchases of this book, please
contact the publishers at the above address.

Contents

Acknowledgements   6

Introduction   7

The Vision   9

How to use this book   11

1 Why cut your car use?   12

2 Travelling less   20

3 Making better use of the car   32

4 What are the alternatives?   48

5 Why are you travelling?   76

6 How much will you save?   90

7 Making it work   108

References   122

Directory   132

Distance, speed & fuel conversion tables   150

Feedback & updates   151

Journey diary   152

Maps   156

Index   159

Acknowledgements

THE GREENHOUSE EFFECT

GOSH, I WONDER WHY WE'RE GETTING THIS WEIRD WEATHER?

This book was written for environmental and safety reasons, and is dedicated to Nature.

I am extremely grateful for advice on draft versions from Harriet Bachrach, Hugh Barwell, Blaise Davies, Guy Jillings, Daniel Johnson, Kate McMahon, Kate Milligan, Anthea Nicholson, Jim Semlyen, Kath Tierney and Jonathan Tyler. Most cartoons are by Andrew B. Singer. The map of the National Cycle Network is from Sustrans, and maps of the main rail network and London Underground are from the London Transport Museum. The extract from the Safer Routes to Schools map on page 81 was drawn by Philip Griffin, and is reproduced courtesy of Radstock Primary School, Earley Town Council, Wokingham District Council and the Pedestrians Association. Thanks also to the Road Traffic Reduction Campaign for their signpost logo on page 12.

John Elford of Green Books and Valerie Belsey also helped with research, and their astute suggestions have made the book more complete.

Introduction

It is with great joy that I offer this self-help book on personal transport.

Who is it for?

- Anyone making travel choices. British adults make around three journeys daily.[1]

- People wanting to limit their driving, save money, be healthier and greener.

- Staff, teachers and/or students involved in environmental studies or reducing traffic.

- Those who have recently moved, as a reference to local transport and amenities.

- People thinking of sharing or selling a car.

The problem is that traffic volumes are large and growing, choking Britain in pollution, harming physical and mental health, cutting off communities and warming the world. Many families have become dependent on the car, but this is unsustainable and increasingly frustrating.

The market for transport does not reflect real costs, because revenue from road users is less than half the indirect costs (such as injuries, air pollution, oil spills etc.) that motoring imposes on society.[2] The variable (i.e. extra) costs per mile of car use are around 57% of the full costs to the motorist[3] and few drivers consider their inefficient use of scarce fossil fuels: most of the energy moves the vehicle's weight, not people or goods. Consequently, cars are driven far more than is necessary or desirable.

Many people think that there is too much motorised traffic. A survey of over 18,000 people in Edinburgh in 1999-2000 found that 88% wanted more money spent on alternatives to the car.[4] However, the Government is not aiming to cut traffic levels. Effective deterrents, such as road pricing or workplace parking charges, which would help the poor by generating funds for alternatives, will be piloted by several councils in the near future. But such measures will not be widely implemented until at least 2005.[5]

Decisions have been devolved to local Councillors who are likely to be wary of upsetting the 72% of car owning households (i.e. voters).[6] It is therefore up to individuals to drive less if they want action. This is a smart move as fuel prices and congestion are likely to rise, making car travel increasingly expensive and time consuming.

Besides, transport is not usually an end, but a means – valuable only to give an opportunity to get to work, to the shops, to school or college, or for another purpose. Cutting your car use will save you money, improve your health and raise everyone's quality of life.

Fortunately there is a lot that individuals can do to tackle car dependency. This book gives detailed practical advice on cutting traffic by changing personal habits. A 'package' of measures will work best. Stay awake to some lifestyle changes such as localising by using the nearest alternatives, and mixing your travel modes.

If every Devon driver used his or her car just two miles less per week for a year then we would save:
45 million car miles
1,495,000 gallons of fuel
10,000 tonnes of air pollutants.[7]

The Vision

I believe that it is possible to reduce traffic. Imagine a land where every child can walk or cycle to school in safety, where companies benefit from green commuting, where local shops thrive and a car is not essential to enjoy life. This fits with the aims of Local Agenda 21 as regards sustainability, i.e. improving the quality of life today whilst ensuring that future generations enjoy a quality of life that is at least as good.

Surveys suggest that 33%–39%[1,2] of drivers would like to use their cars less. Picture how different your neighbourhood would be with fewer moving cars. Safer, quieter and sweeter in many ways. For you personally, how much better would your life be if you were in a car less often? Imagine that improved peace of mind, health, finances, relationships and quality time are yours today.

Don't put up with growing traffic and its problems. Our actions are not just 'drops in the ocean'. Each one of us is powerful if we decide to make positive changes.

This book aims to stimulate awareness of, interest in, desire for, and actions to reduce traffic. Firstly, you must believe you can make a difference, at least to your own travel.

Give yourself six months to learn the full range of travel choices. Begin to use them now for a better future. Persevere, and good luck!

How to use this book

This book lists many ideas for more sustainable personal travel. Only some will fit your circumstances and personality. Apologies in advance for sometimes stating the obvious, or repeating advice which seems important, under several topic headings. I focus on land transport and do not cover the air or water options.

Begin by experimenting with travelling less or using alternatives. It is difficult to change overnight – careful restructuring of your lifestyle takes time.

Organisations in **bold print** are listed in the directory on pages 132-149. As many telephone numbers you need are local, there is space to write them in. Do this soon to make this book more useful. It is pocket-sized to be portable, like a diary, or to be kept by the phone.

• Instructions for positive change are bulleted and indented.

References are numbered and given on pages 122-131. All example costs are dated, as costs rise with inflation.

Chapter 1
Why cut your car use?

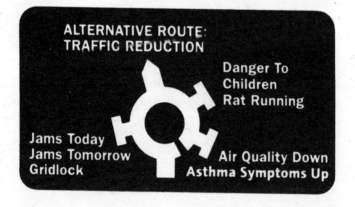

WHY CUT TRAFFIC?

Cars are useful in some situations and essential in others. Successful traffic reduction involves limiting the total number of miles driven, cutting unnecessary trips, more walking and cycling, together with increased vehicle occupancy. The 'green' driver also tries to minimise the environmental and social damage which results from driving.

Even if you are someone who could not manage without a car, or who does not want a car-free lifestyle, there are many reasons for driving less. To get straight to the right section of this book, ask yourself why you want to cut traffic.

To Save Money – see pages 13-14 and Chapter 6 (pages 90-107).

For Better Health – see pages 14-15.

To Be Green – see pages 15-17.

To Manage Time – see pages 17-18.

For a Better Quality of Life – read the whole book and put into practice the proposals.

SAVE MONEY

On average, 15% of household spending is on transport, almost as much as on food or housing.[1] You can add up your own car costs using the charts on pages 94-100. Most cars cost over £50 a week (1999). Could you spend less, earn less and improve your quality of life by cutting your car use? If you travel under 8,000 miles a year, being car-free is cheaper[2] (6,730 miles is a British person's average yearly travel, whilst each private car averages 8,330 miles per year).[3]

• Practise efficient journey planning and linking of tasks on route.

• Act on the energy efficiency advice on pages 40-43.

• Walking is cheap, and saves money otherwise spent at a gym or for health treatments.

• Cycling has low usage costs. Folding bikes can link to a lift, train, bus or taxi.

• Share lifts to save variable motoring costs of 20-25p a mile (1999).[4] Variable costs are petrol, mileage-based depreciation, oil, spares and replacements.

• Use delivery to save on variable car costs, taxis, and public transport, and to save time.

• Make money by working as you travel e.g. by reading on a train.

• Get a weekly or monthly travel pass using discount cards. Book seats well in advance to make use of special offers.

• Small cars (up to 1100cc) pay lower road tax: £55 less per year (1999).[5] From March 2001 the figure goes up to 1200cc.

• Call freephone (0800/0500) taxi or travel numbers.

• Ask for loyalty deals with car hire firms, or join a car club.

BE HEALTHY

Driving is stressful and inactive, and car users breathe the most polluted air: air pollution inside a car is at least twice as high as on the pavement or for cyclists.[1] Cutting your car use will benefit your mental and physical health.

• Move by your own power for a leaner body, and a calmer mind and spirit. Regularly walk or cycle, and thereby reduce risks of coronary heart disease by up to 50% and stroke by 66%.[2] Here's a good example: around 60% of pupils cycle to the Kesgrave High School in Suffolk. Pupils are fit and win at many local sports.[3]

• Take a bus as you'll breathe less pollution: on average 66% the rate of car users.[1]

An estimated 60% of British men and 70% of women are not regularly active enough for good health.[2] The recommended minimum exercise for adults is 30 minutes moderate activity most days. For children it is one hour daily,[2] yet over half of all girls aged 10-16 and over a third of boys do not even take the equivalent physical activity of ten minutes' brisk walk daily.[2] 56% of those aged 16-24 years are overweight or obese.[2]

When you have to drive, good posture requires your back to be supported and shoulders relaxed. Do not sit too far away from the steering wheel.

Further advice is from the **Health Information Line, Department of Health** and **Health Development Agency;** also the **British Lung Foundation, National Asthma Campaign** and **Air Pollution Inquiries.**

BE GREEN

Car culture is unsustainable. Drivers cause casualties, grid-lock, filthy air and social costs, as well as damaging the planet. And 'cheap' oil supplies may only last to 2040 or so.[1]

Road transport is the fastest growing source of greenhouse gas emissions, which cause climate change. Global warming is causing more natural disasters.[2] To save the delicate climatic balance, on average, the industrialised nations may have to cut carbon dioxide (CO_2) emissions by up to 90%.[2] The domestic aim in New Labour's manifesto (1997) was for a 20% reduction of all greenhouse emissions by 2010. Since then, the government agreed at Kyoto to reduce greenhouse gas emissions by 2012 to 12.5% below 1990 levels.

From 1980 to 1996 there was a 63% increase in motor traffic in Great Britain, almost all of which was due to cars.[3] Forecasts for 2016 are that there will be 24%-51% more traffic than 1996 due to increasing goods vehicle use, a four million population increase, and multi-car house-holds.[3] Predictions also show increasing mileage per car and more dispersed activities.[4]

On average, men travel much further than women: 9000 miles compared to 6000 miles a year (1995/7).[5]

Here's what you can do:

• Travel less, localise and buy local products and services. If it is practical, move closer to your most essential and frequent activities. Live close to a town centre to be less car-dependent and reduce urban sprawl.

• Leave your car at home as often as you can.

• Follow the energy efficiency and air quality advice on pages 40-45.

• Work out whether you could use a smaller, more fuel-efficient car or cleaner fuel – see pages 45-46.

• Ask the **ETA** for its free software to calculate how green your lifestyle is. The Ecocalc (ecological calculation) disc is £2 (1999) from **Going for Green**, or ask **Future Forests** how many new trees would offset your CO_2 consumption – on average five per year, costing £20 to plant (2000).

• Join a green breakdown service e.g. the **ETA**, as other motoring assistance groups e.g. the **AA** and **RAC** are involved in lobbying for more roads and cheaper motoring.

• Share a car with another family or join a car club, instead of sole ownership – pages 68-70.

• Write letters to your local MP, Councillors or media to support traffic reduction.

• Support at least one green transport charity – see the Directory.

• Comment on local planning applications for land uses generating vehicle trips.

• Sell one of your household's cars: see pages 101-104.

Helplines include **Going for Green, Are You Doing Your Bit?** and the **Energy Saving Trust**.

Driving is the second most ecologically damaging thing most people do. Air travel is the worst.[6] The greenhouse warming effect of one unit of aviation fuel is three times that of fuel burnt by terrestrial modes of transport.[7]

• Try to restrict your flying to essential trips that cannot be done by land or sea.

MANAGE TIME

Traffic jams are common, and journey times are rising in urban areas. Britons spend an average of an hour a day travelling.[1] Maintenance, parking and walking to get a ticket also take time. So instead of trying to shave seconds off each trip, value time by whether it is spent usefully. Driving is stressful because inattention could be fatal. If you weren't driving, you could give more quality time to your thoughts, to friends and to your surroundings.

• Localise by working, shopping and socialising from home, or nearby.

• Shop and get information by post, phone, fax or internet.

• Combine journeys by doing two or more tasks and linking trips.

• Ask for flexi-time, as then being slightly early or late is not an issue.

• Cycle for speed. It beats a car in urban peak hours and buses for journeys up to 8 miles.[2] Get a folding bike. Some fold in 15 seconds and all avoid parking problems.

• If you are a parent, arrange lift sharing, or buy a teenager a public transport pass to cut your 'taxi' duty.

• Use public transport or a taxi, and work or rest as you travel. Use express train services with good connections and a taxi link up.

• Sell a car, and spend less time working to earn money.

• Drive safely: speeding doesn't save time – particularly on congested, stop-go urban roads.

SUCCESS STORIES

Analyst programmer **Jonathan Powell** from Wantage knew he'd never sell his car. But when it was written off ten years ago he didn't replace it. His family of two adults and three children do more than 85% fewer car miles. Environmental concerns, saving money and less worry or stress from car journeys were motivating factors. He says he is better off, fitter through cycling and has raised his quality of life as every journey is different. "We occasionally hire newer cars than we could otherwise afford."

Chapter 2
Travelling less

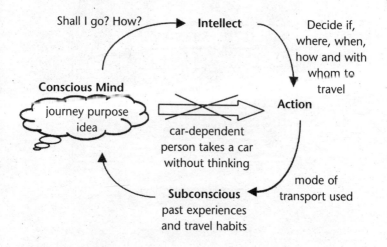

Fig. 1 Making a travel decision

STOP AND THINK

How do people make travel decisions? Figure 1 shows a simple model in which my mind and intellect work in a cyclic pattern to determine how I behave. Imagine a journey purpose idea comes to mind from your subconscious, e.g. shopping. A car-dependent person will think no further about how to travel – only driving comes to mind.

To break free of negative habits, stop and think (rather than act on impulse):

• Where are the people or things involved in achieving the purpose for which I'm travelling?

• Do I really need to travel? • If so, how far? • Is there a nearer alternative? • When shall I go?

• Can I link the trip with another purpose? • How shall I travel?

• Is there public transport? • Is it at the right time? • Can I rearrange to fit timetables?

• Can I share the ride or vehicle costs?

Figure 2 (page 24) shows a flow of decisions about food shopping.

Having asked questions, the intellect judges the evidence and then decides on an appropriate action. All past actions are remembered by the subconscious. The more often (or recently) I use a 'green' mode of travel, the greater its chance of coming to mind. If we do something for at least three days in a row it becomes a habit.

Habits are created by practice, and can be changed!

SUCCESS STORIES

Adrian Lane, a civil servant from Sheffield, sold his car. He says: "We should be prepared to work harder at making the alternatives work – researching timetables, cycling, walking etc.. Remember why you are doing it. For me it was because of information on the consequences of car use, and for health, environmental and cost reasons." Adrian says he is healthier, fitter and happier. His urban journey times are generally shorter, he has more energy, plus new friends and knowledge about the environment and transport issues generally.

STAY STILL

Eliminate unnecessary trips and stay still more often, to save creative energy and quality time.

• Use letters, phone, fax or home delivery instead of travelling. Consult **Royal Mail**, **Yellow Pages**, **Thompson Directories** or **Directory Enquiries**. **BT** Phone Base suits heavy users. To block junk mail and calls, register with the **Mailing** and **Telephone Preference Services**.

• Invite people to your home/office or a middle point, instead of going to them. Offer a map or good directions. This reduces your travel, though not distances overall. Trips for two or more involve three types of movement: $|\rightarrow|$, $|\leftarrow|$, or $|\rightarrow\leftarrow|$. It is efficient overall to meet in the area where most people already are.

• Combine journeys by travelling only for two or more purposes.

• Do all your car-based jobs on one day of the week.

• Consider getting a computer, a modem, and a free email account.

• Ask for flexi-time and to work from home one day a week/fortnight. Distance travelled fell by 73% on teleworking days in a trial.[1] The Royal Bank of Scotland save on travel and subsistence costs by teleconferencing. Ask the **ISI Business Infoline**, **Telework Telecottage and Telecentre Association** or **Home Office Partnership (HOP)**. Home production includes child care, gardening etc..

• Learn a home exercise like yoga (contact the **British Wheel of Yoga**) or tai chi.

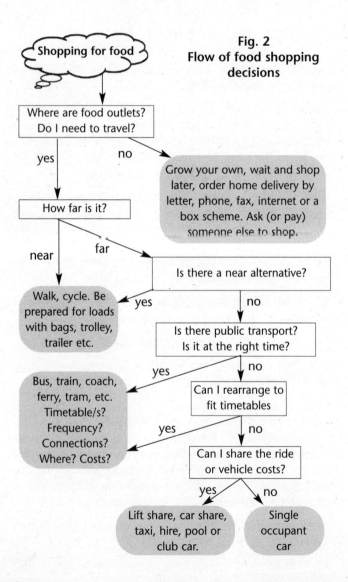

**Fig. 2
Flow of food shopping
decisions**

LOCALISE
Choose the closest options

'Location, location, location' is a catchphrase for house purchase, and it applies to all activities when aiming to cut car use. The point is to reduce the need to travel. The Government can help in the long run by granting planning for high density, mixed use developments. But we each choose how we live today. So don't be distanced . . .

• Localise to save time, money, the earth's resources and because it is convenient.

• Live near an urban centre to be less car-dependent,[1] ideally in a traditional walking city (e.g. York, Durham and Chester), where people live close to each other and there are mixed land uses, or in a public transport city which is medium density, mixed use, grid-based and centralised.[2]

In planning a journey, ask yourself: "Could I achieve my aims closer to where I am, or along my route?"

• Work, shop, study and socialise locally to enjoy more spontaneous exchanges with people.

• Look for adjacent goods and services close to where you are for other reasons, e.g. nearest post offices, corner shops, play areas, cinemas etc..

• Don't drive to take exercise! Walk or cycle more often.

• Don't drive to find the nearest parking space. Walk a little and save time.

• If distances are too far, consider working from home, asking for flexi-time, going part-time, changing jobs or moving.

THE PSYCHOLOGY
OF CHANGING TRAVEL MODES

Habit and the first mode used influence an individual's transport choices for the rest of the day. Dawn decisions are crucial, so in Finland, breakfast products carry advertisements encouraging healthy ways to get to work. Try changing how you get around for your first trip(s) which are:

1. To .... Mode used.........................
Best alternative ....
2. To .... Mode used.........................
Best alternative ....
3. To .... Mode used.........................
Best alternative ....

• Plan the alternatives for regular first journeys. What mode, route, time, who with, etc.? Also get accessories ready, e.g. put your walking shoes by the door with full weather wear. Regular cyclists keep lights/reflectives and a lock handy and have a well serviced bike. Do you have change for a bus or a travel card ready?

Being well prepared with knowledge and equipment will raise confidence about the comfort, speed and reliability of each available travel mode, and their interconnections.

• Put notices up for the morning (e.g. on the fridge, inside the door) reminding you to walk / cycle / take the bus / lift share etc.. Use affirmations like "I make healthy travel choices." Put up pictures of your family travelling wisely, smiling, looking and feeling good.

• Where possible, shorten your first trip by choosing local facilities, work or schools.

Disincentives to walk or cycle are being in a hurry, night time and having a lot to carry.[1] To travel more sustainably:

• Allow time to go by foot or cycle or don't aim to fit in too much, particularly on a first trip. For instance set your morning alarm a little earlier and perhaps cut the least productive journeys (or other activities) from your life.

• Try to arrange travel in the day. At night, use well lit routes and take a pocket torch. Be seen in bright clothes and/or reflectives – see personal safety (pages 109-110).

SUCCESS STORIES

Swindon housewife **Ruth Chivers** drives 50% less to get fit, save money, reduce pollution and be an example to her children. She organises her time so that walking, cycling or using public transport are real choices, rather than driving "in a last minute rush". Doing a few jobs in one round trip makes journeys more productive. Ruth shares lifts with other parents, and talks about car use with her family of four so that it is not taken for granted. "Whatever the weather it is always possible to wrap up warm or wear waterproofs. My daughter has enjoyed the two mile walk to kindergarten since the age of three. Walking, cycling or using public transport is quality time – an opportunity for the family to talk and enjoy the changing seasons. Plus it is quicker and less hassle to cycle to town than drive and park."

• Minimise what you carry, particularly on the first trip. Use lockers, delivery or professional load carrying aids – see loads (pages 113-115).

Some people deliberately make choosing car travel harder to cut their dependency, e.g.:

• Garaging or parking the car at a good walking distance from the front door.

• Sharing a car or lending it to insured drivers to limit availability / temptation.

• Hanging up car keys rather than keeping them in a bag or pocket.

• Deciding not to drive at all on certain days of the week.

• Saving up all car-related errands for an allotted day of the week.

MAPS & DIRECTIONS

We know the way to places in our everyday lives, but maps and/or good directions are crucial for new or infrequent trips.

• Ask for directions before going, or if lost. Carry the address and/or phone numbers.

• Put a local map and the latest cycle route plan up as posters. Ask your Council, **London Cycling Campaign**, **Sustrans**, **Stirling Surveys**, **CityCycle Guides** or at tourist centres.

• Make and give out a map/directions to help people find your home or business.

Always include a map, directions and public transport details when inviting others to travel. An eight digit grid reference is also useful. The easting comes first, and is read horizontally. The northing is last and is vertical.

• Buy the local **Ordnance Survey** 'Explorer' map to find quiet routes and places of interest. Maps are also available from **A–Z**, **Collins**, **Stanfords**, **Bartholomew's**, or **David & Charles**.

CD-ROMS let you print out. **OS** and **Bartholomew's** are at 1:200,000 scale. Some electronic organisers, e.g. Psion, include 'StreetPlanner' navigation aids.

• Consider internet route plans e.g. **RAC** for roads (free). **BT** TelMe charges a fee.

• Ask local bus, train, underground, ferry companies etc. to post you maps, e.g. with price zones and how to get to the nearest network stop (see pages 61-66). Know how to reach your destination from the station you will arrive at, or use a taxi.

SUCCESS STORIES

Colin Newman, an administrator from Barking, uses a bicycle to carry shopping, a folding bike, public transport and the internet instead of driving, because of environmental concerns and cost. Being car-free involves "no parking hassles, no car care, lower costs and a low personal nitrogen, carbon and sulphurous oxides pollution output". Colin has "friends with cars that I could call on" if he ever needs to car share.

TIMETABLES

Unless service frequency is very high, an understanding of timetables is the best way to make the most of public transport. Telephone and internet timetable information is growing, yet it is still handy to refer to printed copies. Tables are usually read vertically, not across. Most use the 24 hour clock (e.g. 3pm = 1500 hours). Watch for special notes, e.g. except Sundays.

• Call your nearest transport companies for timetable information. See pages 61-66.

• Keep tables and notable times together in a folder to be easy to find, cross-reference and choose which public transport options are relevant.

• Recycle out old timetables. Then you won't read last year's.

• Make a note of suitable and last services. Regular interval times are easiest to memorise e.g. departs $1/2$ hourly, at 10 & 40 mins past, 0610–1940, then hourly to 2340.

• Schedule or rearrange meeting times to fit timetables.

• To save time, use express services where possible.

• Work out the interchange time between connecting services. Ideally, use services with good connections that give you time to change (e.g. between platforms).

• Ask operators for easy to read, current and all connecting timetables to be displayed at your local network nodes, e.g. bus stops and stations. Brighton and Hove Metro Bus and Coach company's free timetables have route plans, a

frequency box, and well designed colour presentation.[1]
They won 'Bus Operator of the Year' and Marketing prizes
in the 1998 Bus Industry Awards. Poor timetables have
irregular intervals, poor connections and unmemorable
times, plus wildly different service patterns at weekends
and holidays as compared to weekdays.

SUCCESS STORIES

Engineer **Willy Hoedeman** drives 4,000 miles p.a.
fewer by adapting his life style to principles of appropriate transport. His motivation was financial, after a
series of expensive repairs following vandalism, and
crash damage. He and his family have ensured that
they don't live too far from work or school. They use
bikes for commuting, to get around town, and for most
shopping trips. Willy also rides in the country, and does
some fitness training; in all, he cycles about 50 miles
weekly. He uses his classic Volvo for special holiday
treats, and a VW Camper for occasional weekend trips
to the country and for recycling journeys. It is also useful for moving his Woodcraft Folk youth group. The
annual total mileage for both vehicles is under 4,000
p.a. Willy commutes faster by bike, is in better health
and has higher self-esteem than if he had driven. He
keeps his vehicles parked or garaged away from home
so that they are not immediately accessible and are less
tempting. Willy recommends others to "try to move
closer to work and school, and seriously invest in the
alternatives, such as a rail card and a good bike, and to
start cycling for fitness, fun and enjoyment".

Chapter 3

Making better use of the car

SLOW DOWN

Lower speeds are crucial to limiting the impact of traffic: excessive and inappropriate speed contributes to one third of all crashes.[1]

Why is speed so important? In physics, $e = \frac{1}{2} mv^2$. The energy of an object [e] hitting a car or person is half the mass [m] or weight of a car times its velocity [v] squared (i.e. speed). Even quite small increases in speed increase risk dramatically.[2]

Survival and Speed[3]

Vehicle Speed	<10 mph	10-20 mph	20-30 mph	30-40 mph	60 mph
Pedestrian/Cyclist's Survival Chances	very good	95%	55%	15%	almost no chance

Safety and emission control are better at lower speeds (except at very low speeds).[2] Slowing down reduces fear in people on the street (especially children), improves accessibility and benefits health.[3]

69% of car drivers in Britain routinely break the 30 mph urban limit.[3] Do they save time? Research on three-mile trips across York in stop-go driving conditions found only a 20 second difference between going at 30 and 25 mph.[4] What use is 20 seconds?

Opinion polls support slower speeds. 80% of Carlton TV viewers favoured a 20 mph limit on residential roads in London,[5] and in a poll of 1200 parents, 90% wanted 20 mph to be the maximum limit outside schools.[5]

• Reduce your maximum speeds, as every 1 mph less reduces accidents by between 3% and 6%.[6] Drive smoothly, and slow down in adverse conditions, e.g. fog, rain or darkness. Save yourself speeding fines and penalties.

• Be especially cautious if driving in urban areas or villages where there are likely to be pedestrians and cyclists.[2]

• Reduce your rural road speed, as 70% of fatal car crashes and 50% of cycle deaths happen on rural roads.[1]

Slowing down traffic is possible. Between 1990 and 1997, using a combination of speed cameras and advertising, the state of Victoria in Australia cut the percentage of vehicles exceeding the speed limit from 23% to under 2%.[7]

SUCCESS STORIES

KW belongs to Leicester's Rusty Car Pool, so called because its first car was an old banger. Now they have 22 members and five or six vehicles, including estates, hatchbacks and a van which converts into a minibus. For KW the group is "mainly about sharing resources because our little terraced streets have too many cars, not everyone needs one and they take the space where kids would have played." A car pool suits those who do not drive everyday and want to share an expensive resource or live on a moderate income. As well as cutting costs and the number of cars on the streets, other pooling benefits are that each vehicle is used more often and efficiently than a private car plus the extra flexibility to use the type of vehicle best suited to the journey's purpose.

ROAD SAFETY

10 people die a day,[1] and almost third of a million people a year are hurt on British roads: there is a 1 in 200 chance of being killed on the road in your lifetime.[2] The lifetime probability of a child born in the 1990s being seriously injured in a road crash is 1 in 4.[3] The annual cost of road death and serious injuries is at least £15,960 million (1998):[4] road deaths alone cost 2% of Gross Domestic Product.[5]

Follow the Highway Code, and be particularly careful at junctions, roundabouts, bends, and in darkness. Mutual respect for all road users is vital to safety.

Driver Safety
• Drivers must be insured and all car users must wear seat belts where fitted.

• Reduce speeds, be patient and limit overtaking. Never overtake on a bend.

• Don't drink and drive. Nearly one in seven road deaths involves drivers who are over the UK limit (80mg of alcohol per 100ml of blood [1999]).[7] Since any alcohol impairs driving, it is safer not to drink at all.

• Ask a doctor or pharmacist about driving whilst on medication, e.g. antidepressants, painkillers, eyedrops, insulin, antihistamines, anti-diabetic or epileptic drugs.[8]

• Take rest breaks on long journeys.[9]

• Never use a hand-held phone while driving. It is safer not to use a hands-free phone either.[10] It is the distraction of the conversation, not the mechanics of using the phone, that is unsafe.[11] Use a telephone message service instead.

- Have eye checks at least every two years.

- Consider extra tuition e.g. **Pass-Plus** or **IAM** (£95 1999).

Vehicle Safety
- Use the smallest, lightest vehicle available for the purpose, to minimise danger and damage done by driving.

- Regularly maintain and test brakes, check wear on tyres, tyre pressure, lights, pedals etc.. Position the seat and mirrors correctly.

- Don't have bull bars. The front of vehicles should have a design that causes minimum impact damage to pedestrians and cyclists.[12]

Safety from Others
- If you need to stop on a motorway hard shoulder, wait on the embankment.

- If you are flagged down and unsure if it is a genuine emergency, don't open a window or get out. Indicate that you will drive to find a telephone to get help.

- Park in well-lit busy areas. Keep doors locked and valuables out of view.

Environmental Safety
- Enhance visibility by having a bright coloured vehicle, good lights, reflective strips and clothing for cyclists and walkers plus reflective collars for pets. The **Technicolour Tyre Co** supply reflectives by mail order .

- Do not block entrances or emergency access when parking.

• Observe parking restrictions at road junctions e.g. 20 metres for 30 mph zones.[12]

• Ask your Council's Road Safety Officer ☎.................... or Police Road Patrol Officer ☎........................ for leaflets/training, slower speeds and home zones.

Ring the Police about any road crime. **RoadPeace** and **Brake** advise victims. Other safety groups are **RoSPA**, the **Slower Speeds Initiative**, **Child Accident Prevention Trust**, **Children's Play Council**, **PACTS (Parliamentary Advisory Council on Transport Safety)** and **CPRE**.

Injured cyclists can call **CASCA** for free 'no win, no fee' legal advice, or get legal advice and 3rd party insurance through **CTC** or **Bycycle** club membership.

SUCCESS STORIES

Retired educational psychologist **John Taylor**, from the rural village of Corpusty in Norfolk, has reduced the cars owned by his household (of three adults) from three to one. His son no longer drives; he now walks and uses buses instead. John is worried by the transport crisis of too much traffic, especially lorries, and aims to raise safety for cyclists and walkers. He remembers a time when cars were not so misused as today. John's advice is to "rethink how you travel and make a shopping list so that you don't forget things". John doesn't accept lifts and is happy to walk, even in bad weather. He sees the benefits as being healthier, saving money and being a good role model.

TRAFFIC FLOW

Do you get stuck in traffic? 1.6 billion hours were lost on British roads because of congestion in 1996; 80% in urban areas.[1] Jams are caused by drivers' toleration of hold-ups before some eventually decide not to travel, or switch modes or times.[2] Gridlock is the extreme, where only walkers and two-wheelers can move. Peak traffic consists of commuters, business trips and the school run in term time. Improve traffic flow by driving shorter distances, less often, and . . .

• Live in a town, localise, combine trip purposes, use delivery and flexi-time.

SUCCESS STORIES

L. Nash and **B. Brett** from Hereford have two cars but both cycle to work daily and walk to the shops, carrying loads with a trolley. This has cut their driving by about 50 miles a week. They wanted to save money and the "hassle of getting around, as there are very bad jams most days at peak times". Getting rid of their cars is not yet an option as one works as a stonemason and they occasionally move heavy stones and pull trailers. Their advice to others is to find another way of getting to and from work. Benefits are of getting around the locality faster, with less stress and they are fitter. Most importantly, they make "large savings on fuel, spares and depreciation".

• Ask yourself: "Is the car really necessary?" One in four car journeys is less than two miles,[3] and cars are being used for shorter and shorter journeys.[3]

• Leave plenty of time and drive mostly at off-peak times.

• Plan routes carefully with a map. The **RAC** gives free route advice at <www.rac.co.uk>. Or buy software such as Microsoft Auto Route Express.

• Before going, check out internet sites such as <www.rac.co.uk> or <www.nadics.org.uk> for traffic advice.

• Look for roadwork warnings in local media and listen to radio traffic bulletins.

• In-car devices warning of congestion ahead include the **RAC**'s Traffic Alert (free to new members) or **AA**'s Personal Roadwatch 1800 (£30 in 1999).

• Lift share, and use high occupancy vehicle (HOV) lanes, where they exist.

PARKING

There were 27.5 million vehicles taxed for use on Britain's roads by the end of 1998, of which over 23 million were cars.[1] Vehicle stocks are so high that space is running out, both on the roads and for parking. Road congestion and parking are the most unreliable aspects of any car trip. So if you have to drive . . .

• Ask for good directions, a parking map, car park pass or about charges.

• Consider using a Park & Ride if available.

• Many areas have resident or staff parking schemes. Find out if you need a permit, and keep a stock of visitor permits ☎ ....

• Allow extra time if you anticipate parking problems.

• Carry change for meters or pay and display, out of sight in the car.

• Park responsibly in well-lit, designated areas keeping emergency/path access.

• Do not mount pavements, as this damages surfaces and hinders walkers.

• Observe parking restrictions at junctions e.g. 20 metres in 30 mph zones.[2]

• Consider renting out surplus parking space or garages, or converting them. A parking space is worth £770 per year (1999) in central Leeds.[3] Buildings with off-street parking cost more and are subject to higher council tax. Could you re-locate, and thereby save on estate costs? The Spring 2000 Transport Bill, if enacted, will give local authorities powers to levy workplace parking charges.

ENERGY EFFICIENCY

Energy used for transport increased by 48% due to a 63% increase in road traffic between 1980 and 1996.[1] Cutting energy use saves money and fumes.

• Drive less, particularly for short trips, as a cold car engine produces 60% more fumes[2] and uses more fuel than when warm.

• Use the smallest, lightest vehicle available for the purpose. Check fuel use before buying as similar models of cars can vary by over 25%.[3] The **DETR** and **Vehicle Certification Agency** produce a guide to new car fuel consumption and carbon dioxide emissions twice yearly.

• Record miles per gallon or 10 litres. See the Conversion Tables on page 150.

• Plan journeys in advance. Start the engine only when ready to go, and set off immediately. Avoid revving up, and push in a manual choke as soon as possible.

• Limit speeds: 40-55 mph is the most economical speed.[4] At 70 mph you use up to 30% more fuel than at 50mph.[2] Drive steadily and read the road to manoeuvre smoothly. Heavy feet wear out brake pads and tyres.

• Get in the right gear, preferably top. Change before 2,500 revs per minute.[2]

• Avoid idling. Shut off the engine when standing still for one minute or more.[4]

• Have a catalytic converter to cut emissions of carbon monoxide, hydrocarbons and nitrous oxides when travelling over five miles. Unfortunately converters do nothing to cut emissions of carbon dioxide (the chief global warming gas), are usually only guaranteed for 50,000 miles, and use scarce precious metals.[4]

• Know that automatic transmission can add 10-15% to fuel use, and that air conditioning uses an average 15% more fuel.[3]

• Consider which type of fuel to choose: diesel is 30% more fuel efficient than petrol, but has worse health impacts.[4] 'Clean fuels' can be 50% more fuel efficient than petrol.[5] See pages 45-46.

Maintenance
• Check tyres monthly. A 7psi under-inflation wastes half a gallon per tank.[3]

• Have your car serviced at least every year, or each 10,000 miles, to ensure the engine is properly tuned. Get the emissions and the catalytic converter checked. 90% of badly polluting vehicles can be retuned at a garage within 15 minutes.[2]

SUCCESS STORIES

Information analyst **Gabriel Hughes** sold his car because of the cost and hassle. It needed repair and MOT and he "couldn't be bothered". Moving to West Hampstead, London from Kent made it easier. Gabriel regularly travels by train to Brighton to see his fiancée and occasionally shares her small, fairly new Skoda on country trips. "My housemate has a fancy car and is into the status thing but spends a lot of time in traffic jams." Gabriel is saving money, and is reassuringly "never stuck without a car – you can always get a taxi in an emergency and in the long run that is not so expensive." He says to "investigate ways that trains, tubes and buses fit together or you might miss a very quick route from A to B just because it is not immediately obvious e.g. from a tube map."

- Consider buying a fuel efficiency device. An **Ecoflow** magnet can save 10%-20% on fuel – saving £120 a year for a one-off cost of £50 (1999). *Permaculture Magazine* claims toxic fumes are cut by half.[6]

- Streamlining kits and aerodynamic styling cut fuel bills.

- Call the **Energy Saving Trust** for free energy packs.

Are You Doing Your Bit?, **Going for Green** and the **Alternative Technology Association** also advise. The **Energy Efficiency Best Practice Programme** advises companies.

AIR QUALITY

Exhaust fumes are poisons: they damage lungs and hasten climatic change. Road traffic produces one-fifth of carbon dioxide,[1] over half of nitrogen dioxide and over 75% of carbon monoxide emissions in the UK.[2]

Nearly one in five people in the UK is at risk from dangerous traffic pollution levels.[2] One in seven children has asthma, a problem triggered by exhaust fumes.[2] Up to 24,000 people per year in the UK die prematurely due to road pollution – more than in crashes.[1] A similar number are admitted to hospital due to respiratory problems from air pollution.[2] When pollution is high, those with heart or lung disease may have to modify their treatment and avoid strenuous outdoor activity, especially in the afternoons.[3] If traffic fumes make breathing harder, try to avoid busy streets.[4]

In-car air can be three times as toxic as air breathed by cyclists or walkers,[5] as cars are in the middle of the road, where pollution is most concentrated. Bus users face, on average, a third less pollution than car users.[5]

Diesel fumes are the most damaging to health, due to tiny particulates called PM10s. The health cost to the community of diesel has been estimated at 84p per litre of fuel (at 1993 prices).[6] Traffic's external costs to society are three times those paid in road taxes.[6]

Summer smog is worst on hot sunny days. Winter smog occurs in urban areas on calm days, often after a cloudless night and a morning ground frost or mist.[2]

• Smell the air. Look and listen for air quality warnings in local media. If worried, ring your Council's Air Quality officer ☎ .... or contact **Air Pollution Inquiries** for updates/literature. Try not to drive at all in smog.

• Act on this book's energy efficiency advice (see pages 10-43), for example by limiting speeds.

• Try not to use diesel, because of the particulates. See cleaner fuels on pages 45-46.

• Have your emissions checked. Roadside spot checks are made by Government Examiners. Seven local authorities can also make checks, with £60 spot penalty fines for high emissions (1998).[7]

• In traffic congestion, adjust the ventilation so that it does not draw in fumes from the vehicle in front.

• Report smoky vehicles (heavy goods vehicles, buses and coaches) to your **Vehicle Inspectorate** Enforcement Office, giving the number plate details. Contact the **DETR**'s helpline.

• Ask the **DETR**, **Department of Health** or **Health Information Service** for free leaflets.

As well as acid rain, one of the effects that air pollution has had on the natural environment is a reduction in the amount of light that reaches the earth's surface: the Smithsonian Institution has reported a 14% loss of light intensity in a 60 year period. Reduced sunlight is believed to increase crop diseases and failures.[8]

Charities and campaigns include the **National Asthma Campaign**, **British Lung Foundation**, **National Society for Clean Air**, **Stop Fuming Campaign**, **Local Government Association**, **Friends of the Earth** and **Greenpeace**.

CLEANER FUELS

All cars pollute, and advancing technology will not produce a clean car for several years, if ever.[1]

• To improve air quality, consider switching to one of the three 'clean fuels': liquefied petroleum gas (LPG), compressed natural gas (CNG) or electricity.[2]

Hybrid cars can use two fuels e.g. gas and petrol. Gaseous fuels burn cleaner, which reduces piston and cylinder wear. They use less oil, and fuel use is reduced by up to 50%.[3] Electric vehicles are for short, urban journeys.[3] 'Runabouts' should be lightweight electric or hybrid-electric with a 25 mph top speed.[4] The government is subsidising alternative fuel vehicles with grants of up to 75% of the extra purchase cost.[2] Ask the **Energy Saving Trust** about Powershift. Or the **Liquid Petroleum**, **Natural Gas** or **Electric Vehicle Associations** or **Alternative Vehicle Technology**.

- Catalysts cut emissions after five miles, and are guaranteed for 50,000 miles.[1]

- Leaded petrol has been replaced with unleaded, lead replacements or additives.[5]

- City diesel is cleaner than ultra low sulphur diesel.

- Service the car regularly, and consider a fuel efficiency device such as Ecoflow.

Get the **National Society for Clean Air**'s free guide *Clean Cars: how to choose one* or the **ETA**'s *Car Buyer's Guide*, judged on environmental damage.

QUIET

Listen. What can you hear? Traffic – or maybe car alarms? Noise from road traffic and roadworks is forcing people to live in their back rooms. Between 1983/4 and 1995/6, complaints to Environmental Health Officers about traffic noise increased by 83%.[1] Traffic disturbance depends on the quantity, volume, speed and route. Vibrations are also particularly influenced by weight. So . . .

- Use the lightest vehicle available for your aim (from shoes upwards).

- Buy products and services locally.

- Drive below the speed limit. 20 mph is more appropriate than 30 mph in towns.

- Get the engine tuned, and do not rev it unnecessarily.

- Avoid 'rat runs' in residential areas or country lanes.

• Turn down in-car music, and use the horn only if there is no quieter alternative.

• Do not slam vehicle doors, set off car alarms or leave them sounding.

• Report noise nuisances to your Environmental Health Dept ☎ .... or to the **DETR**'s Vehicle Standards or Aviation Divisions.[2]

• If many people live on a busy road, ask the Council for porous asphalt surfaces to cut traffic noise.[3] Get the **DETR**'s free leaflet on insulation against traffic noise[2] or consider relocating to a quieter street. The **National Society for Clean Air** also advise.

Hearing is an important sense, and reasonable quiet is needed to talk and listen to each other. We also all need to sleep soundly to cut stress.

Chapter 4
What are the alternatives?

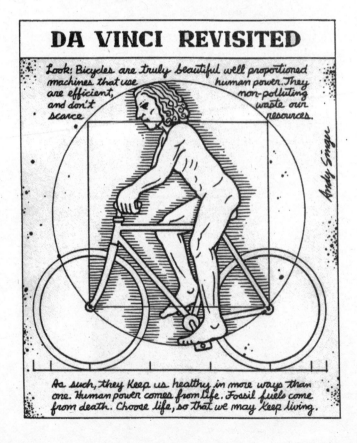

DA VINCI REVISITED

Look: Bicycles are truly beautiful well proportioned machines that use human power. They are efficient, and don't waste our resources. non-polluting

Andy Singer

As such, they keep us healthy in more ways than one. Human power comes from life. Fossil fuels come from death. Choose life, so that we may keep living.

WALKING

Go green on foot! Walkers endure three times less fumes than people in a car;[1] they can safely chat, and meet friends and others en route. Walking just over a mile a day significantly reduces heart disease risks, and is better than low salt diets for lowering blood pressure.[2] It increases bone mass and reduces fat.[2] Go at 4 mph and make savings with the cheapest travel mode.[3]

• Wear shoes/boots that fit well with flat soles, good grip and ankle support.

• Get weatherproofs: coat, bag, hat, scarf and gloves.

• Be seen with reflective/high visibility clothes/strips or sew strips to a bag, e.g. from the **Technicolour Tyre Co**.

• Have route knowledge/a detailed map/directions and use safe road crossings. Ask your Council for extra safe crossings where needed ☎ ....

• Dropped kerbs/ramps/tactile surfaces are useful for the disabled or pram users. Ask your Council's Walking Officer ☎ .... for an access guide. **The National Federation of Shopmobility** loan equipment. Advice on carrying loads is on pages 113-115.

• Use a stick/crutches/frame or push a trolley if unsteady.

• Report broken pavements or lighting to your Council's maintenance department ☎ ....

Charities include the **Pedestrians Association**, the **Ramblers Association** and **Help the Aged**. Personal safety advice is on pages 109-110.

LIFT SHARE / RIDE SHARE

Lift sharing uses spare vehicle capacity, saves money and provides social company. Ride sharing is when lifts become regular. 20-25p per mile is saved in variable costs by not driving (1999).[1] For a 10-mile round trip, 220 days p.a. (i.e. 2,200 miles), savings are £220–£275 p.a. each if taking turns to drive. Three sharers save 40–50p a mile, £290–£370 each p.a. Passengers also save on parking fees and time. Sharing taxis saves even more. If a car is sold, perhaps £2,000 is saved per year.[2]

I believe that sharers will increasingly be rewarded with faster travel in high occupancy vehicle (HOV) lanes on roads such as the A647, preferential parking spaces and cheaper parking passes.

SUCCESS STORIES

Marcus Jones lives near his transport consultancy work, to which he commutes by bicycle. Didcot has a station and he is within walking distance of shops and public transport. Marcus and his wife sold their car as they didn't like driving or wasting money and believe that unnecessary car use is socially irresponsible. They bought good luggage carrying equipment for their bikes and a shopping trolley. On rare occasions when a car is needed they hire the right vehicle for the job. Marcus advises others to "add up the true costs of driving before choosing to live far away from work". As a non-car owner he enjoys savings and greater freedom.

• Ask (or email) family, friends, work mates or other parents to share lifts.

• Put up lift share notices and register with **National CarShare**, **London Car Share** or **Freewheelers**.

• Ask your firm for a transport notice board and map.

Lifts Wanted/Offered – template for a noticeboard

From	To	Depart approx.	Return approx.	Days/ Dates	Spaces	Name	Phone
		...am/pm	...am/pm				

Those interested in sharing could mark their home on a map. Lift share databases e.g. **Entec** can match postcodes, as can graphical information systems e.g. **MapInfo**. It is recommended that potential sharers meet informally before. I could share with:

Name ....☎.................. Email ....

Name ....☎.................. Email ....

Name ....☎.................. Email ....

• Always meet in a public and well lit place.

• Passengers should ask the driver for the make, model and colour of the car as well as the registration number. Don't get in the car if these do not correspond.

• Passengers should give someone else details of who they are meeting, where, and when, where they are going and what time they expect to arrive at the destination.

• If working shifts, ask to be on the same shift as others you could share with.

Savings from regular ride sharing can pay for taxis in exceptional circumstances. Organised commuter ride sharing schemes are best backed up with a guaranteed taxi home, if sharing temporarily breaks down. This is tax-free for employers.[3]

• At stations, or before leaving an event, ask others waiting for a taxi if they are going in your direction.

Hitch-hiking is an informal kind of lift sharing. See page 73.

TAXIS

A taxi is a 24-hour, chauffeur-driven car at your door with very little notice. They are convenient and cut parking time and worries about directions. Journey times are quicker as taxis can use bus lanes. Taxis are potentially high occupancy, require less road space than if everyone had their own car and provide jobs. Crucially, a taxi is safer if you plan to drink – you may otherwise lose your licence, or worse.

• Try **Yellow Pages** or **Thomson Directories** for freephone numbers (0500, 0800). Call firms near home/work about fares. The best have in-car computer systems to allocate jobs geographically.

....☎.......................................
....☎.......................................
....☎.......................................
....☎.......................................

• Put up a notice about regular taxi share e.g. to town from a village to buy food.

Virgin run a Carlink scheme to book a taxi to or from a rail station. Or phone **National Cabline** if your don't know a local firm.

York's standard taxi rates in 1999 were: first 535 metres or 2 mins 37 seconds £1.70. Each succeeding 153 meters or 45 seconds 10p[1]. Within the ring road, a taxi is about £3.50 to a supermarket. You could take two local taxis a day instead of owning a car. Taxis cost less than a standard car hire for around the first 30 miles per journey.

CYCLING

Cycle for independence and health. At 10–20 mph from door to door, It is the fastest travel mode in urban peak hours.[1] Cyclists breathe less than half the fumes of people in a car,[2] and enjoy more reliable journey times.

Cyclists can reach an area 12 times larger than the catchment accessible by foot[3] as bikes are three to four times faster for a fifth of the energy.[4] Non-exercisers who began cycling around 19 miles weekly rapidly improved their aerobic fitness by over 11% in a trial.[5] Body fat can also be significantly reduced[5] by this low-cost exercise

which is easy to incorporate into daily life.

Cycling five miles, four times a week, can cut risks of coronary heart disease by up to half.[6] Regular activity also reduces chances of strokes, colon cancers and osteoporosis.[6] Indeed, the health benefits of cycling outweigh its injury risk according to the British Medical Association.[7] Cycling is clean, quiet and poses little risk to anyone else.

Pavement cycling is illegal, though children under 16 cannot be fined for doing it.[8] Anyone can mount the pavement if forced to.

Folding bikes can fit on a train, bus, coach, taxi or ferry. You can accept a lift or drive a stage before cycling on. Folding bikes give you more flexibility than using cycle racks on cars. See *A to B* or *Encycleopedia*. **Brompton**, **Cresswell** and **Moulton** are British makes, or try **Birdy**, **BMW**, **Strida** or **Di Blasi**. Disability cycling is expanding. Try **Chevron Handcycles**, **Bromakin**, **London Recumbents** or **Companion Cycling**. Also see child trailers on page 117.

• Get a bicycle/folder/tricycle/recumbent/tandem/unicycle for your gender, height, weight, ability and purpose. Buy modern gears and a lightweight machine for an easier ride. Adjust it to your ideal riding posture. Hire if you are unsure what to buy. The **CTC** website has a list of hire firms.

• Get lights for night riding complying with the British Standard 'Kite mark'. Dynamos are greener and cheaper than batteries in the long run.

• Buy a good lock – e.g. D lock and chain. See cycle security on page 59.

• Put reflectors on the bike and wear reflective clothing for visibility. Reflectives are available from **The Technicolour Tyre Co.**

• Wear a comfy jacket, trousers, gloves, hat etc.. Ideally these should be waterproof.

• Get a cycle friendly map with gradients and an optional map carrier. Avoid major roads, difficult junctions or right turns. Use **Sustrans'** National Cycle Network (map on page 156), **CycleCity** Guides, **London Cycling Campaign** maps, or the National Byway maps (from the **AA**).

SUCCESS STORIES

Simon Collings, an environmental consultant from Didcot, moved to within five miles of his job and began cycle commuting so that he could sell his car. Simon was motivated to become a one (rather than two) car household to reduce expenditure, be more environmentally and socially responsible and for the health and stress relieving benefits of cycling instead of motoring. He has cut his car use by a total of around 3,000 miles a year, as he walks and cycles more on other trips too. Access to a good rail link 15 minutes' walk away means that he and his wife use trains for shopping or leisure trips to Oxford and Reading. He says, "Start cutting your car use by examining your current journey patterns and seeing where you can use alternatives. Start by making small changes. Ask for advice from others who travel by the mode you are thinking of changing to about the best routes, and how to overcome the barriers you perceive to using that mode. Also, when moving home, think about the travel implications."

• A cycle helmet is optional, but is recommended to protect against head injury.[9]

• Consider insurance for personal accident, injury and/or theft, e.g. **CTC, Bycycle Club, London** or **Cambridge Cycle Campaigns**. The **Compensation and Advice Service for Cycling Accidents (CASCA)** is free.

• Take a water bottle/flask on longer trips.

• Ring your Council's cycling officer for maps and training ☎...................:... or **Cycle Training**.

• Both employees and the self-employed may claim tax relief of 12p per cycle mile (1999), or on the difference between their employer's allowance and 12p, if the former is lower.[10]

• Children and other loads can be safely carried on a bicycle. See pages 113-117.

• Write to the **Cycle Campaign Network** for local contacts.

Checks Before Riding
• Are tyres in good condition and inflated to the pressure shown on the tyre?

• Can you rock the front of the bike by the handlebars with the front brake on without movement of the headset (i.e. the bolt your handlebar stem slots into)?

• Are brakes and gears working efficiently with cables that are not frayed?

• Are the saddle and handlebars adjusted to the correct height?

General Maintenance
• Clean and lubricate the chain with Super Spray or Finish Line (not WD-40).

• Check chain tension. It should be firm, not sloppy.

Bicycling Books stock books on all aspects of cycling.

BUYING A 'GREEN' BICYCLE

As a mode of transport, cycling is about as ethical as you can get. But choosing an environmental bike raises dilemmas: for example in regard to some of the materials (aluminium and titanium) used in its manufacture. Aluminium, for example, is typically mined as bauxite in Brazil, worked in Sweden then sent to Japan to become tubing. It then travels to be made into handlebars in the USA, to Taiwan to be fitted to a bike, and finally to the USA or Europe for sale![1]

Most European and US cycle companies (which are usually marketers and distributors, rather than manufacturers) have relatively clean environmental records, except those importing from China. It is always better to buy from a local independent cycle shop than a national chain. Using a second-hand bike also lessens your eco-impact. Other factors to consider are:

• Whether the company is local and materials are transported long distances.

• Whether the company has a history of environmental harm or irresponsible marketing.

- Whether your purchase supports high street bike shops, rather than large sheds on the by-pass.

- Whether the company operates in oppressive regimes or the arms trade.

- Whether the people who made the bike have good working conditions and rights.

- Whether the company contributes to cycle campaigning.

SUCCESS STORIES

Mark Eastgate from Huddersfield, an environmental consultant, deliberately walks, cycles or takes the bus instead of driving. He aims to protect the quality of life of children of the future, rather than perpetuate congestion, stress and pollution. Mark goes 6,000 car miles per year fewer and only drives once every three weeks to pick up his daughter in Oldham. His small car is driven at a top speed of 50 mph for fuel efficiency, and he stops to pick up hitch hikers to fill spare seats. "I would never contemplate driving the 1.5 miles to town any more, as short car journeys are worst for pollution because the engine has not reached its proper operating temperature." Mark tried to share his car, but his friends are mostly cyclists and don't trust his old banger. His life has become more measured, and Mark says "I laugh to myself when I see people getting stressed in their cars".

CYCLE SECURITY

As there were more than six thefts per 100 owners in 1993,[1] cycle security really matters. Larger BMX and mountain bikes are stolen twice as often as other types.[1] Unusual, ladies or visually unattractive models face less risk.

• Note the bike's make, colour and frame number. Consider insurance e.g. adding a bike to a household policy or through the **CTC**.

• Lock a bike to an immovable object. 37% of bikes stolen in 1993 were locked and only 5% were stolen from locked premises.[1] Use a **Bikepark**, if available.

Only 8% of thefts involve cracking a D or U lock.[1] Crime Prevention Officers say that if all cyclists used high quality D-Locks (costing £30+), theft could fall by 50-90%.[2] 'Sold Secure Ltd' test and endorse locks. Look for their logo.

• Have two bikes and/or disguise quality with dark paint or take a folder inside.

• Postcode or tag your bike. One third are coded and 17% are returned if coded compared to 13% if uncoded.[1] Die stamping on the bottom bracket, or UV pen marker kits are offered by many police stations. Register with the **National Cycle Register** or fit an electronic tag.

• Fit anchor points for short term parking in your yard or shed e.g. the Abus Wall Anchor, ring-ended Rawbolts, a wall bar or Sheffield stand – ask the **CTC**.

ELECTRIC BICYCLES

Hills and headwinds are largely overcome with power-assisted cycles, either purpose built or as power kits added to a normal bike. Electric bikes are exempt from tax, insurance and MOT if they fulfil the UK Electrically Assisted Pedal Cycle regulations (1983). This requires a motor to give a maximum 200 watts. Current is automatically cut off if the rider stops pedalling, or rides at less than a medium walking pace of under 2.2mph (3.5km/h). It also cuts off above 15mph (24km/h).[1] Minimum age is 14.

A friction-drive kit is an easily installed, low cost unit, but is affected by rain. A purpose-built hub drive is most efficient, quiet and reliable, but spares and servicing cost more. Chain-driven rear wheel or pedal crankshaft bikes are purpose built and efficient. Folders and tricycles are also available. The average is 15 miles per charge.

• Ask for free brochures from **CAT Electric**, **Easybike**, **Electric Bike Co**, **Raleigh**, **Sinclair Research**, **TGA Electric Leisure**, the **Electric Vehicle Company**, or look at bike magazine adverts, e.g. in *On Your Bike*.

• Buy a good battery. Lead-acid dry-cells are cheap, easy to maintain and reliable. Nickel Cadmium holds 25% more energy by weight, so you can go further. But they are costly and output gradually diminishes if not serviced.[1]

BUS / PARK & RIDE

A typical bus can seat 50 people.[1] Passengers breathe a third less pollution than in a car[2] and priority measures are making trips quicker. Let a professional drive as it is at least twice as safe[3] as driving yourself, with no parking hassle. Buses are usually clean and comfortable. New ones have low floors and cleaner fuels. Plus, public transport is often cheaper than driving, even for two adults.[4] Half-price fares for pensioners are coming in and there are weekly and daily passes. In Ireland, OAPs travel free on all forms of public transport.

Some firms have 'real time' information on displays at stops. National telephone and internet bus timetables are being set up.

☎.................................www...

☎.................................www...

• List local bus and Park & Ride telephone numbers.

Check your Council's bus line ☎........................... or bus planner site www...
or try **UK Public Information**, **Journey Call**, **London Transport** or **Tripscape.** Local bus companies and nearest stops are:

....☎...

....☎...

• Ring for timetables, fares, passes or joint ticket possibilities, or about the location of stops.

• Write for a free copy of the **Royal Mail** Postbus passenger timetable.

Plan your bus trip, knowing the departure time or frequency. When and where is your ride home? If the last public service is too early or late, solutions include taxi, lift sharing, staying overnight, taking a folding bike or walking. Certain rural bus companies and disability services operate a flexible service. Passengers ring a control centre to ask it to stop at their home.

• If needed, ask Social Services about disability transport
...........................☎..
...........................☎..

See the advice on personal safety and loads on pages 109-110 and 113-117. While waiting for a bus:

• Read the signs/timetable or ask others waiting to check it is the right stop.

• If possible, choose a stop that is well-lit and overlooked for safety. A seat, timetable, shelter and being near a public phone are all desirable.

• Have your pass or change ready.

• Know that you are cutting pollution and saving money compared to using a car, but carry taxi money and phone numbers just in case.

• If you are unhappy with services, write to the operator and then to the **Bus Appeals Body**.

Your employer could help to raise bus use and cut parking pressures.

• Ask for discounted travel passes, business meetings and shift times to match timetables.

• Join or set up a Bus User Group to push for incentives.

TRAINS & UNDERGROUND

Despite high profile disasters, rail travel is statistically very safe[1] and can be faster than a car over long distances or in congested areas. Plus, you can use time on a train to relax or work. Value this time at your hourly wage.

UK prices favour return journeys. Peak times are usually up to some time before 10am, 4–7pm, and on Fridays. Increases on key fares are restricted to match inflation. Example fares and discounts, with price ratios against an open return and pence per mile are given below. All discount fares are less than marginal car costs of 20-25p a mile (1999).[2]

York–London 377 miles return.[3] (Dec 1999 rounded fares)	Fare £	Price Ratio	pence per mile
First Class	177	1.58	47
Standard open return (any train)	112	1.00	30
Saver (after 8.15am out)	63	0.56	17
Supersaver (after 10am, not Fri or back 15.30-19.00)	53	0.47	14
Super Advance (buy before 6.00pm the day before travel)	45	0.40	12
Apex (buy 7 days in advance, book trains)	38	0.34	10
GNER bargain (7 days in advance, book trains)	31	0.28	8
Daypex (1 day in advance, return same day)	26	0.23	7

• Buy tickets well in advance for better availability of discount seats.

• Go early. **LTS** give 20-30% discounts to those arriving in London before 7am.

There are also railcard, season and holiday deals.

• Get 33% off leisure fares and All Zone One Day Travelcards with a Young Persons Railcard (£18 p.a. 1999), plus discounts in participating stores. It is for 16-25 year olds and students attending lessons 15 hours weekly, 20 weeks p.a.

• Get 33% off most fares with a Senior Railcard (£18 p.a. 1999) for 60+ year olds.

• Get 33% off 1st and standard fares with a Disabled Railcard (£14 p.a. 1999).

• Get 20% off leisure fares and £2 child fares with a Family Railcard (£20 p.a.).

SUCCESS STORIES

Rosemary Turner used to live near Newhaven Harbour and need a car. She then moved to Brighton. When someone wrote off her car, she bought a computer with the insurance. This momentous decision opened a more satisfying career as a freelance editor and charity administrator. Rosemary is now true to her values and only does work she approves of. By downshifting "I am free to do the work I love without worrying about overheads." Rosemary finds walking healthy and meditative. She likes observing the characters on the bus and mixing with different types of people, as she is a writer who otherwise works alone. A taxi helps bring the weekly shop for herself and her daughter up a steep hill.

• Ask for group discounts e.g. 10+ and 'through' tickets, e.g. including buses.

You may break the return trip with a 1st class, standard open, saver or supersaver return. 'Smart cards' and 'carnets' for pre or electronic payment are coming in. Certain fares are linked to underground, buses, taxis or car hire. A few centres (e.g. **CAT**) offer discounts to non car-borne visitors.

• Look on a map for nearest rail stations and work out how to get there and back.

The GB Main Railways Map is on page 157, and the Central London Underground on page 158.

• Pick up timetables and brochures or book a ticket at a station or phone and ask for information to be mailed.

• Ring **Rail Enquiries** ☎08457 484950 (24 hours), **London Transport**, **Eurostar**, **Rail Europe Direct** or the train operator. Book with the train operator using a credit/debit/charge card. Tickets are mailed or picked up at a selected station.

• Surf the timetable and/or book online at <www.railtrack.co.uk/travel>, <www.thetrainline.com> or <www.pti.org.uk>, or buy **Rail Planner** software.

• Customers with special needs can book in advance for help at stations.

• Reserve space for a non-folding bicycle, which usually costs £3 (1999) per trip.

• Frequent rail users or businesses may want to buy the twice yearly Great Britain Rail Passenger Timetable £9 (1999)[3] and/or a fares manual £13.50 (1999)[4].

• Keep valuables and luggage in sight.

Unlike being caught in a traffic jam, you can claim a refund for a serious train delay. If you are unhappy with services, write to the train operating company. The appropriate **Rail Users Consultative Committee** helps those dissatisfied with the response. Ring the **British Transport Police** to report any incidents.

COACH

Coaches are a safe[1] way to travel, and cost less than standard rail tickets, but the trip can take longer. **National Express** serves over 1,200 UK destinations ☎08705 808080.

Save with a coachcard for £8 p.a., or £19 for three years (2000): available to students, 16-25 and 50+ year olds. Family Saver (£15 p.a.) Coachcards enable two children free travel with two adults. LoneParent Coachcards (£8 p.a.) let you take one accompanied child free, except on **Eurolines UK** to main European destinations. Tourist Trail Passes give 2, 5, 7 or 14 days unlimited travel. However, coachcards or passes are not valid for trips entirely within Scotland on **Scottish Citylink**.[2]

Stagecoach Oxford run a 24 hour service to London. ☎ 01865 772250 <www.stagecoach-oxford.co.uk>.

• List local coach firms and ring for timetables, or try **Journey Call**. Other local coach companies and nearest stops are:

....☎...
....☎...

HIRE CARS

If a vehicle is occasionally needed for longer journeys, car hire could be the answer. Small car hire in York is £27 a day, £55 for a weekend, or £170 a week; this includes insurance, but excludes the first £250 worth of damage (1999).[2] You can get eight weeks' hire, or 26 weekends, for less than the fixed costs of owning a small used car.[3] Car hire and mixing modes will be cheaper than car ownership for you if you travel less than 8,000 miles p.a.[4]

Choose British Vehicle Rental and Leasing Association members, who offer arbitration to customers. Local car hire firms are:

....................................☎....................................
....................................☎....................................
....................................☎....................................

• Ring for brochures and offers.

• Look for deals integrating car hire with other tickets e.g. rail etc..

• Check whether firms provide baby and/or booster seats for toddlers if required.

• Register with a hire firm for a (10% plus) loyalty discount and fast track service.

• Comprehensive insurance is costly. Consider taking the first £250 risk.

CAR SHARE

Private cars are unused most of the time. If driven for an hour each weekday, it is still unused for 96% of weekdays and 100% of weekends, though parked in different places. Sharing car ownership can save a great deal, £1,000+ per year each[1] for two people (1999). See pages 95-97.

• Ask friends, colleagues or put up notices about shared car ownership. Try to find local people whose needs 'dove-tail' against yours. I could car share with:

..................................☎...

..................................☎...

Whilst car share can be informal, someone must be responsible for servicing, maintenance and insurance.

• Write a simple agreement between all sharers to build trust and respect.

Split fixed costs according to time used, and running (variable) costs by mileage. Agree how the car will be booked, where it is parked, and what happens if different sharers want it at the same time. One person may have priority on certain days of the week, or whoever books it first. Normal hire is a back up! Agree how the sharing arrangement will end, if someone decides they want to leave it.

• Record miles in a logbook. Sharers usually pay for their own petrol, so agree to fill the tank before swapping over, or note petrol costs and miles driven.

The **Community Car Share Network**, **Smart Moves'** *Car Club Kit Book* or **Local Exchange Trading Schemes** (LETS) can help.

CAR POOLS

Car pools are shared cars, usually leased by a firm, but which may also be borrowed for personal use by staff. Advantages are of a flexible, cost-effective capital resource compared to individual, non-essential company cars in cases where regularly hiring cars would cost more. Car pools take up less parking space than private cars.[1]

• Ask for car pools at work. Buy small, fuel-efficient model(s). Hire or take a taxi if you occasionally need a larger vehicle.

• Take care of insurance/maintenance/servicing and **Inland Revenue** issues. Tax-free allowances are higher for the first 4,000 miles (1999).[2] See page 94.

• Agree rules for business and non-work time sharing. Pools can be staff car clubs. e.g. instead of a parking space, staff get free hire for three weekends per year.

• Use a booking system e.g. a diary or wall chart/stickers.
☎ ....

• Log miles driven in a book in the car. Consider a mileage charging system, and the rules for who pays for petrol. Staff must keep records of allowances for **Inland Revenue** returns.

• Consider having dedicated parking space(s) near the main entrance.

• Get the **DETR**'s free guide *Companies and Cars: The Way Forward.*[3]

Smart Moves and the **Community Car Share Network** can also advise.

CAR CLUBS

There are car clubs in over 250 European cities, including **Edinburgh,** Leicester and Leeds. Other clubs include Cranfield University, and there are pilot schemes in Bristol and Earlsdon. Members pay a subscription. They then ring to book for preferential car hire rates. Loan is by the hour, day or week. Mobility Car Sharing is popular in Switzerland with 29,000 members (Aug 1999).[1]

One club car replaces 4–5 owned cars, increasing space for amenities, people and public transport.[2] Many former owners reduce their overall car use by 50%.[2] Savings are around £1,400 (1998) per year over a mid range car – more if your mileage reduces.[2]

• Contact your council, the **Community Car Share Network** or **Smart Moves** to find, or start, a local car club. **Smart Moves** sell the *Car Club Kit Book* (£7.50). My local car club booking number is: ☎ ....

• Car club management advice is available from the **Lothian and Edinburgh Environmental Partnership** (LEEP).

MOPED, SCOOTER, MOTORBIKE & TRIKE

Motorcyclists have the highest casualty rate at around 16 times that of a car user.[1] Safety concerns prevent anyone from recommending two wheeled powered vehicles other than electric bicycles. Motorcycles can be noisy, but they use less road space than cars.

Smaller and electric motorbikes pay less road tax. 1999 rates for motorcycles not over 150cc were £15, 151cc-250cc £40, above 250cc £60 p.a. Electric motorcycles are only £15 to tax per year, whilst tricycles cost £15 up to 150cc and all others are £60 p.a.[2]

Basic safety advice in the Highway Code is overleaf.

SUCCESS STORIES

Jemima Jefferson lives in Leytonstone and works in the City of London. She has been riding a motorbike for twenty years and this is what she has to say.

"If you ride a bike to save time getting to work but you have the mentality of a car driver it isn't long before you break a leg in several places. You must learn riding skills before you venture out into traffic on a bike. It seems that it's almost impossible to learn them when you have been driving a car for a long time – it's a different mind set.

"The beauty of motorbikes is that they are a true alternative – they have a whole lifestyle that goes with them. You can park them all over London on motorcycle parking bays that cost nothing, they are easy to maintain, cheap on fuel and kinder on the environment than cars in queues.

"I like my motorbike – it's fun, quick to get through traffic, exciting on a sunny summer Sunday, and for me, pretty safe. You can't drink and drive if you ride a bike because the risks are too high. The statistics show that old bikers are unlikely to go head first through car windscreens or wrap themselves round trees."

• Rider and pillion passenger on a motorcycle, scooter or moped must wear a protective helmet. Helmets must comply with regulations and be fastened.

• Eye protectors which conform to regulations are advisable.

• Consider wearing ear protection.

• Strong boots, gloves and clothing may help protect you if you fall.

• Not more than one pillion passenger can be carried and they must sit astride on a proper seat and keep both feet on the foot-rests.

• A white helmet and fluorescent clothing or strips will help visibility

• Dipped headlights, even in daylight, may also make you more conspicuous.

OTHER ALTERNATIVES

Trams, such as the **Stagecoach Supertram** in Sheffield and **Tramtrack Croydon** operate in some towns, much like buses, but on fixed tracks. My local tram company number is: ☎ ....

Roller blading The roller blading action is like ice skating. It is said that it does not take much practice. Blading is quicker than walking, though not as fast as cycling. It is greener and healthier than car use because blading uses human power.

Blades are of two types, 'recreational' and 'aggressive'. Choose recreational blades (longer than aggressives and with a heel brake) which are faster and harder wearing. Wear pads for knees and elbows and consider a wrist guard. Show respect to pedestrians.

Blades cost £40-£200 (1999).[1]

Roller skating is harder than blading – and not as fashionable! Quad skates have four wheels in a square.

Skate boarding, even when just pushing with one foot on the flat, is two to three times faster than walking. Boards are very light, and fit easily under your arm when not in use. Utility skaters need good bearings with big, soft wheels, and to wear protective pads to lessen injuries.

Respect pedestrians; use the pavement or a cycle route, not roads with motor traffic. Use carefully chosen routes according to surface quality, as kerbs slow you down.

Hitch-hiking is cheap, green and often a pleasant adventure, but I cannot recommend solo hitch-hiking: it is best done in pairs for personal safety. It uses spare vehicle capacity, and rides are free. Picking up hitchers makes for quicker journey times if there are high occupancy vehicle lanes, such as on the A647.

Hitching suits a minority, but is the least reliable mode. It works at Lancaster University which has hitching posts for lifts to town. **Freewheelers** will advise you.

If you do hitch-hike, pack sustenance and waterproofs; plan routes with a map by major road junctions and service stations. Consider how to get to the most sensible start point to hitch (e.g. the nearest junction, roundabout or service station exit on your route). Have money and a back-up plan or maps, as you are unlikely to get a direct lift.

The Highway Code says you must not pick up or set down anyone, or walk, on a motorway, except in an emergency; or stop on a Clearway.[2]

Horse/pony Before fossil fuels, British farms had small ponies to provide personal local transport, horse-drawn holidays, muscle power, manure and company. Native pony breeds (e.g. Shetland, Dale, Welsh Cob, Fell Highland etc.) can stay out all year round only requiring extra hay in winter.

However a horse can cost as much to buy and maintain as a car! Ponies are cheaper, and need less room. The younger the animal, the cheaper it will be. A foal can be as little as £50, a yearling £800, and a fully grown, schooled pony £3,000.[3]

Lyn and Chris **Dixon** advise on ponies. Further information is in *Understanding Your Pony*,[4] *Heavy Horse World* or *Native Ponies* magazines. The **British Horse Society** or **Byways and Bridleways Trust** offer advice on horse buying, rider training, stabling and bridleway maps.

JOURNEY PLANNING

Could you reduce your need to travel, link journeys better or mix travel modes more?

• Record a personal diary of all car trips for a week in the tables on pages 152-155. This is highly recommended as it is the most systematic way to look at how you currently travel. Start by filling in the columns you know.

• Plan the alternatives. The information in this chapter on the various modes will help. Remember that you could change vehicles e.g. share a lift out and get public transport back.

Circular trips with more than one stop often cut miles travelled overall. You can break your return journey with some rail tickets e.g. standard open returns, savers or super savers. Consider an overnight stay if there is no convenient service home, e.g. the **British Tourist Authority**, **YHA** or **SYHA**.

• Compare your current car use with the alternatives for each separate journey.

Could you save money and be greener by using more local facilities or changing journey patterns? See page 94 about tax and car allowances and pages 90-101 on travel costs.

SUCCESS STORIES

Project manager **MB** from Oxford drastically cut her car use from 12,000 to 4,000 miles p.a. because of cost, the environment and fitness. She 'commuter ride shares' with four others. Although work involves a 30 mile round trip, she deliberately chose city living, near bus routes and local shops. She drives to a supermarket just once a fortnight, and takes a bus to the station even though her job will pay for a taxi. MB books cheap rail tickets and works on the train – an effective use of company time. Her advice is to cycle. "It's just a habit, gets you fit, is sometimes quicker – you just need the right clothing and accessories." MB could not sell her car – there are no public services to work and she shares the drive. But she saves £600 p.a. in fuel plus £100 in servicing and spares even after transport fares. She is fit enough not to need aerobics or circuits. Also, her car is safer on her drive than in public car parks, and her home looks occupied and is less at risk from burglary.

Chapter 5

Why are you travelling?

COMMUTER JOURNEYS

It matters how you journey to work! Full time jobs take 220 days a year, and if you work from aged 18–60, a lifetime's commuting will comprise over 18,000 single trips. 70% of commuting in Britain is by car,[1] and the average distance of over 12 km in 1993-95[2] had risen by 50% since 1973. Commuter journeys make up 30% of private mileage.[3]

• If possible, choose work by its proximity, or move closer.

• Although society is increasingly 24 hour, try to avoid shift work (unless the job is very close) as public transport is less frequent overnight. Ask to be on the same shifts as others you could share the journey with by foot, cycle, car or taxi.

• Request flexi-time. Then being slightly late or early is no problem and you can fit hours to timetables. Or agree your preferred working times e.g. to do 36 hours, 3 x 12 hour slots or 4 x 9 hours cut travel compared to 5 x 7 hours 12 minutes. Compressed working is when you can take a day off provided hours are worked in advance. After its introduction in 1991 by the City of Irvine in California, there was a 16% reduction in sick leave.[1]

• If possible, work at home at least a day a fortnight and do 10% less commuting. The Royal Bank of Scotland uses home working on a voluntary, casual basis.[1]

Advice is from the **ISI Business Infoline**, **Telework Telecottage and Telecentre Association** or **HOP**.

• Employers can provide cycling kit tax free. Ask for allowances, secure, convenient racks, pool folding bikes, a bike fleet for work/breaks e.g. **Adshel**'s cycle fleet in Cardiff city. Some bike shops hire with maintenance contracts.

• Join or set up a Cycle Opportunities Group (COG). **Sustrans**, the **CTC** or **Cycle Campaign Network** can advise. Or become active in a Bus User Group.

• Consider part-time work. Being car-free may save one day's net wages weekly.

• Talk to managers about Green Travel Plans with the **DETR**'s free guide.[1]

Other Green Transport Plan options include transport information e.g. this book, transport discounts/subsidy, maps, directions, teleworking, teleconferencing, cash instead of parking or a company car, interest-free travel loans, lockers, showers, a car sharing database with preferential parking and taxi backup, on-site facilities e.g. cash machines/nursery/canteen, 'dress down' days and walking initiatives such as zebra crossings and low kerbs.

Green travel brings financial and many other benefits to staff and firms. Advice is from the **DETR**, **Energy Efficiency Best Practice Programme**, **Transport 2000**, **Association for Commuter Transport**, **TravelWise**, **ETA**, **Community Transport Association**, **Sustrans**, the **Ashden Trust** or Anna **Semlyen** – my own consultancy, also called 'Cutting Your Car Use'.

BUSINESS JOURNEYS

Business journeys are those done in the course of work e.g. seeing clients. Individuals generally have less choice over business than commuter travel. So . . .

• Check job specifications for travel levels. Ask if there is a Green Travel Plan.

• Ask about refunds for public transport, taxi fares or discount travel passes. Does the firm use delivery, ideally cycle couriers?

• Ask if there are pool cars, folding bicycles or cycle trailers. The Countryside Agency has a pool of folding bikes for staff to use on site visits after making most of the trip by rail.

• Ask for details of lump sum and mileage allowances for walking/bicycle/car. At least 27p a mile is needed to compensate a fully insured cyclist for costs of around 22p a mile (1999), given that only 12p is tax free.[1] See pages 91-92 for an analysis of these costs.

• Some adverts require applicants to have a car and licence. Don't apply if you don't wish to travel for work, or be honest and say you don't want to. Challenge it if driving is not essential on the basis of 'car transportism' – a kind of discrimination. If a car is necessary, the company should provide one.

• If possible, tailor your work to be close by and carefully plan delivery runs, ideally limiting them to certain days of the week, or geographically.

• Complain if your work schedule means driving for hours with no break, driving hundreds of miles a week, or trying to fit in too many calls a day. **RoSPA** campaigns to increase occupational safety awareness and responsibility.

• Switch to a job with no, or limited, work-related travel or go part-time instead.

SCHOOL JOURNEYS

School journeys are the first journeys in a day; they affect child health and the travel habits of these future adults. British schooling is for 195 days a year for 11-13 years, i.e. 4,000–5,000 trips. Whilst in 1971 80% of seven and eight year olds walked to school, by 1990 it was down to 9%.[1]

School escort trips are an estimated 20% of morning peak traffic[2] adding to pollution and a vicious circle of car use by parents afraid to let children go anywhere alone. A child is 50 times more likely to be killed by a driver than by a stranger[3].

Travel Awareness
• Choose the most local, appropriate school. Re-locate, or work nearer if necessary.

• Teach children the personal safety advice on pages 109-110.

• Choose a quiet route and walk or cycle together. You'll all be healthier. These options are low-cost, non-polluting, socially minded and what youngsters prefer.[4]

• Ask children to draw the locations of hazards on their normal route(s). Suggest safe routes as a classwork topic.

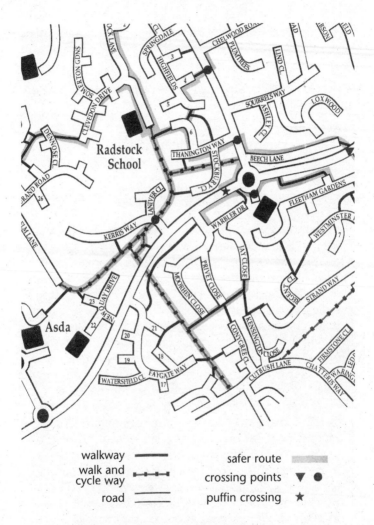

Fig. 3 Safer Routes to Radstock Primary School (section of map)

• Put up local walking, cycling and public transport maps at home and ask for a school journey notice board with maps, timetables, car share etc..

• Ask for a safer routes map in the prospectus and a school policy against car use.

Local Safety
• Ask for slower speeds (20 mph) and safe crossings, e.g. with crossing patrols. Some schools designate dropping-

SUCCESS STORIES

Jo Rathbone from Earlsdon, Coventry, cut her mileage from 915 to 355 miles per month by sharing a car with another family. They each have it on alternate week-ends, with other time negotiated. "Not having a car instantly available is the best way of not using it, especially for short journeys." For Jo, finance was the main reason for driving less, plus green issues. She was happy to not to own a car, but her husband wanted to. "When we heard neighbours were considering buying one and didn't really want to, we teamed up. Sharing works very well." Jo plans her trips carefully, and feels calmer. "A more measured lifestyle is a big plus." They now shop locally and are more active in the community. "We are more dependent on others for lifts, which builds relationships." Jo cycles and walks more, which raises her fitness. "The children love travelling by train and bus and are understanding that we can't always do everything. Hobbies must be in walking/bussing distance." Jo plans to join a car club as soon as one is established.

off points away from school so as to avoid congestion.

• Ask for car-free entrances and to hold cars back until pedestrians and cyclists have left school. Cars leave 15 minutes later at the Copse School, Reading.

• Ask the school to be involved in helping to match lift share partners.

Walking and Cycling
• Be seen, be safe! Sew reflective strips on coats and bags, check cycle reflectors and ask children to wear a high visibility belt, tabard, arm band, cycle clips or hat, helmet etc.. Ask for a review of the visibility of uniform.

• Ask for, or offer to run, an escorted walking scheme and walk to school events. A 'Walking Bus' entails an adult collecting many children along a prearranged route to school. Some have a trolley for carrying bags.

• Keep bikes properly maintained. Buy a regular service if you do not have repair skills. Burnholme Community College in York has a bicycle MOT service run by older pupils to tag cycles as roadworthy (1999).

• Pavements are for pedestrians. On-road cycle training is essential before secondary age. Cycle trainers ☎ ....

• Ask for secure and visible cycle shelters.

• Ask for a homework timetable review to reduce the amount children carry.

• Ask for storage for books, cycling gear and outdoor clothing, e.g. lockers, and a drying room for wet clothes.

Public Transport

• Take your children on public transport. Teach them about timetables and independent travel.

• Ask the school to explore options for new services e.g. better routes and low fares linked to a code of behaviour for school bus users.

Sustrans offer free information packs and newsletters about Safe Routes to Schools.

• Get a free copy of *A Safer Journey to School: A guide to school travel plans for parents, teachers and governors* from the **DfEE**.

Also see pages 116-117 about carrying children.

SUCCESS STORIES

Steve and Carol Fletcher from Twickenham became non-car owners after a crash, in which their car was written off. Steve's computer consultancy contract had ended, and so replacing the car was not a priority. They now enjoy using buses and trains more, as there are no parking problems, they can chat with friends more, and walking to the shops is less trouble, and healthier. The Fletchers use public transport for work, get food home-delivered, and their two children walk to school. Friends lend them their second car at weekends for free, even though Steve and Carol have offered to pay part of the fixed costs. Steve says "there must be many second cars unused at weekends, which could be loaned out." They may get another car, but meanwhile are saving a lot of money, and can always hire a car when really needed.

FOOD SHOPPING JOURNEYS

Over 5 billion car miles a year in Britain are food trips[1]. Home delivery for a market town catchment can save around 70-80% of the driving distance.[1] See Fig. 2 on page 24 for a flow chart of shopping journey choices.

• Perhaps grow some food yourself? Try **Chase Organics** or **Primal Seeds**, or see the **Green Books** catalogue.

• Walk or cycle to find local food shops. Ask what they offer and if they deliver, ideally by cycle courier.

• Are there doorstep deliveries of milk etc.?
............................☎................................

• Share a car or taxi with neighbours. Fix a weekly day and time to share. I could shop with.............................
............................☎................................
• Make a list so you don't forget important items.

By buying direct from farms you'll get the cheapest and best seasonal produce. Organic 2000's box scheme costs 10-20% less than supermarket fruit and vegetables.[1] The **Soil Association** have set up Local Food Links to help consumers buy through:

• Vegetable box schemes. • Food co-ops.
• Farmers' markets.   • Community-owned farms.
• Or ask at wholefood shops for local farm delivery services and stocklists.

My local vegetable box schemes are:
............................☎................................
............................☎................................
............................☎................................

If buying through a store, you could . . .

• Ask for a stocklist/catalogue or check the internet and phone, post, fax or email your order e.g. **Sainsbury's**, **Iceland**, **Tesco**, **Traidcraft** or **Organics Direct**.

• Set up an account and regular shopping list with your favourite food store(s).

• Shop as usual and have it delivered from **Tesco**, **Iceland**, Co-op, Marks & Spencers etc.. Some take phone/fax orders and shop for you. Ideally, use a cycle delivery company such as **Zero Emissions** or **Wheel Alternatives**.

• Check delivery fees. They may be free or less for a certain value of goods.

• Take a taxi to bulk buy and do other trips only possible by car.

• The **National Federation of Shopmobility** loan mobility equipment. Could you get Meals on Wheels? Ask Social Services ☎....................... See pages 113-115 about loads.

My local food stores or delivery firms are:

.................................☎.....................................
Fax:.................... www....................................
Delivery Charges ....
.................................☎.....................................
Fax:.................... www....................................
Delivery Charges ....
.................................☎.....................................
Fax:.................... www....................................
Delivery Charges ....

TAKE-AWAY FOOD JOURNEYS

Avoid driving at meal times just to get something to eat.

• Take a packed meal and water bottle or flask. In good weather find the nearest green space to eat outside. Ask for picnic tables and an indoor staff room.

• Walk or cycle to get something to eat locally or at a work cafeteria. Are there pool bikes that can be used for breaks?

• Use the work kitchen to prepare food or ask for a small cooker, microwave and fridge, or take your own kettle and tea/coffee equipment in.

• Ask a catering firm to visit your workplace regularly or phone for a delivery.

Local catering firms I could get to deliver include . . .

.............................☎..

.............................☎..

.............................☎..

Join, or organise, a work rota to share getting take-aways.

SOCIAL JOURNEYS

We can talk to others in a car, although the driver should always be attentive to safety. Yet, if we see a friend outside, usually the best a car user can do is honk or make a hand sign as they go about a pre-planned aim.

Remember the danger of drinking and driving.

• Meet at your home or in the middle. Post or fax a map and travel information with alternatives to the car on invitations.

• Stay overnight with friends or family, or at the **YHA**.

• Move closer to your family or dearest friends to see them more, provided that this also fits with your work or other regular trips. Choose a street with light traffic. Then you'll have more friends and acquaintances.

• Walk or cycle around to meet people in your community in an unplanned way, cultivating what Greeks called the 'agora', where we meet in the street and during daily activities because everything is close. Walking and cycling provide more opportunities to 'exchange', such as conversation, than other modes per mile travelled.

• Make local friends by using local shops, joining groups or helping neighbours, e.g. by sharing ladders, watering plants, pet sitting etc..

LEISURE JOURNEYS

Country trips are one reason people give for car ownership, but encasing ourselves in polluting, noisy metal boxes is self-defeating. It cuts us off from, and ultimately destroys, the natural beauty, the quality of the landscape and the peace we want to be near. Roads and traffic kill wildlife and divide habitats.

• Take a leisurely walk around your neighbourhood. Explore from your door.

• Walk or cycle to traffic free places. Use the **Sustrans** National Cycle Network (page 156) or the National Byway (maps from the **AA**).

• Some leisure activities can be reached by public transport. The **British Tourist Authority** sells guides.

• Ring for help on getting to e.g. **National Trust**, or **National Park** destinations other than by car. Some attractions offer discounts to non-car borne visitors e.g. **CAT**.

• Try growing vegetables to get you outside and cut food miles. See page 85.

If cities were car-free, we wouldn't need or want to leave as there would be less noise, pollution and danger.

• Ask your council for car-free hours or days of the week e.g. parts of Holyrood Park, Edinburgh are car-free on Sundays and many town centres have part-day driving restrictions.

HIGHWAYS DIVIDE HABITATS

Chapter 6

How much will you save?

MY IDEAL MIX
OF TRANSPORT MODES

What would your household's ideal blend of travel modes (e.g. walk, cycle, delivery, bus, taxi, lifts, train, coach, hire car etc.) be, if you had one car fewer? In other words, you wouldn't have to skimp on taxis or car hire when you need them. Fill in the chart below:

Mix of Transport Modes	Cost £ p.a.	Miles p.a.
Walk		
Cycle		
Car Hire		
Taxi		
Bus		
Train		
TOTAL for mixed travel modes (per year)		

CYCLE COSTS

Consider an adult driver switching to cycling commuting 25 miles per working week. Subtract insurance depreciation etc. if your bike kit costs considerably less than the £500 in the example overleaf.

Fixed Costs per year for a bicycle (1999 prices)
rounded figures[1,2]   **£ p.a.**

Interest lost on capital cost of £500
  (including lights, basket, pannier, etc.)
  minus inflation @ 3%   £15
Annual depreciation (estimated value £80
  after 6 years i.e. £420/6)   £70
3rd party insurance, Legal Aid & personal
  injury insurance (CTC adult)   £25
Total Fixed Costs   **£110**

Running Costs
Servicing and spares (for 1,200 miles,
  varies on use)   £55
Clothing including waterproofs, shoes,
  gloves, hat, scarf, replacements   £40
Theft insurance @ 10 % of value
  (optional & 15% in London)   £50

Totals	**Not Theft Insured**	**Fully Insured**
Total cost per annum	£215	£265
Total cost per week	£4.00	£5.00
Full cost in pence per mile (1200 miles p.a.)	18p	22p
Running cost in pence per mile (1200 miles p.a.)	9p	13p
Income required per mile (23% tax + 10% NICs. 12p tax free, capital allowances unclaimed)	21p	27p

MIXED TRANSPORT MODES AND COSTS

Mixing travel modes is often cheaper than owning a car if you drive less than around 8,000 miles a year.[1] An example for a cyclist is given below (figures rounded to nearest £5). This is shown on a graph on page 94. Bus users will need a pass, say £9.50 per week.[2]

Mix of Transport Modes (Cyclist)	£	Miles
Walk 1 mile more daily (£125 for waterproofs/shoes over 5 years)	25	360
Cycle (£500 purchase price over 6 years)[3]	215	1200
Cycle trailer (£150 purchase price over 5 years)	30	
Home delivery of goods	20	50
Lift share (contribution to petrol)	20	180
Take 2 taxis weekly @ £4 (£1.60/mile, first mile £2.40 then £1)[4]	415	270
Bus fares	25	100
2 saver return rail tickets e.g. York–London @ £62.60	125	755
5 Apex rail returns e.g. York–London @ £38	190	1885
Hire car @ £155 for 1 week[5]	155	
Hire cars @ £50 for 8 weekends[5]	400	
Hire car petrol (84p[6] a litre of fuel, 7.7 miles per litre)	130	1200
TOTAL for mixed modes (1999 prices)	**£1,750**	**6,000**

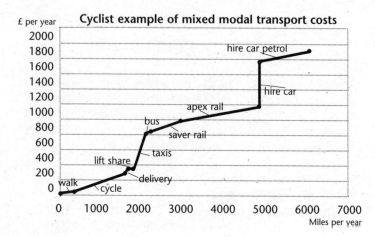

Cyclist example of mixed modal transport costs

CAR ALLOWANCES AND COSTS

The **Inland Revenue** allows you to claim tax relief for business mileage. Tax deductible amounts vary according to the miles travelled. The table below will help you quickly calculate your approximate costs per mile (1999-2000).[1]

Engine capacity	On the first 4,000 miles in the tax year. FULL cost per mile	Excess over 4,000 miles in the tax yr. VARIABLE cost/m
1,000 cc or less	28p	17p
1,001–1,500 cc	35p	20p
1,501–2,000 cc	45p	25p
over 2,000 cc	63p	36p

For low mileage drivers, costs per mile will be much higher than in the tables.

- Work out your actual car costs using pages 95-98.

1. CAR PURCHASE COSTS

Add up your car costs by filling in the blanks with your figures and a calculator.

A My car, with all its extras excluding interest cost   A ....
B My initial payment (cash and/or pt exchange) was B ....
C I could sell my car for (phone a dealer or look in your local paper's car ads section, then deduct at least 10%)   C ....
D I have owned my car for (months)   D.............

Borrowing Costs

Complete E to K if you borrowed money to pay for your car

E I borrowed   E ....
F My loan was for (months)   F ....
G My monthly repayment   G ....
H Total loan cost including interest = F x G   H ....
J Total interest paid = H - E   J ....
K Interest paid each month = J divided by F   K ....

New petrol car costs, 1100-1400cc engine (1999)

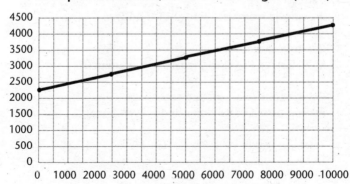

Option 1

I bought my car without borrowing any money

L Each month my car is worth less *(A - C) divided by D*   L..............

M If I had not bought a car, I would have earned interest
   of 3% per year (after inflation) *A x 0.03*   M ....

N My annual purchase cost = *L X 12 + M*   **1N** ....

Option 2

I am still making loan repayments

O My annual purchase costs = *(K + L) x 12*   **1O** ....

Option 3

I bought my car with a loan which is now repaid

P Each month my car is worth less
   = *(A + J - C) divided by D*   P ....

O My annual purchase costs = P x 12   **1Q** ....

2. OTHER FIXED CAR COSTS

I pay these every year no matter how little I drive.

A Car tax   A ....

B Breakdown membership (e.g. **ETA, RAC**)

                                         B ....

C Car insurance   C ....

D MOT test fee (excluding repairs needed to pass!)

                                         D ....

E Garage costs (e.g. rent and council tax)   E ....

F Parking permit costs (home and/or workplace)

                                         F ....

G Total fixed costs = *A + B + C + D + E + F*   **2G** ....

3. RUNNING COSTS

Every time I drive I spend more.

A Miles per year the car is driven   A..............

B Miles per litre of fuel on average
(mpg divided by 4.546 = miles per litre)   B..............

C I buy = *A divided by B* litres of fuel per year   C..............

D Fuel costs per litre (in pence)   D..............

E Total cost of my fuel = *(C x D) divided by 100*   E..............

F 12 months parts, servicing costs, repairs and oil   F..............

G In a year, parking, tolls and car washes cost   G..............

H Driving fines (parking, speeding etc.)   H..............

J Variable running costs before depreciation = *E + F + G + H*

3J ....

K True variable running cost with mileage-related depreciation
of 3.5p^2 (1998) a mile = *J + (0.035 x A)*   **3K**..............

(4. & 5.) YEARLY CAR COSTS

4 Ownership costs

Before I use my car it costs me

A Total purchase costs: Either 1N or 1O or 1Q

A ....

B Mileage-based depreciation of 3.5p a mile (1999)
= 3.5 x 3A divided by 100   B ....

C Total other fixed costs: 2G   C ....

D Total ownership costs = *(A - B) + C*   D ....

5 Total annual costs

A Total annual cost of my car = *2G + 3J + 4A*   **5A** ....

B Total weekly cost of my car = *A divided by 52.18*

B ....

6. INCOME TO PAY FOR A CAR

How much income must you earn to pay for a car?

A My top rate of income tax is   A .... %
B I pay National Insurance Contributions at
                                      B .... %
C My marginal tax rate is = $A + B$ (33% is standard 1999)
                                      C .... %
D The total yearly income to run my car =
    5A divided by (100 - C) x 100   D ....
E   A weekly income = *D divided by 52.18*   **6E** ....

7 Working Hours
A My hourly wage is   A ....
B I work 6E divided by A hours a week to pay for my car
                                    B ....

A cost example is given on pages 99-101, and shown in a graph on page 95.

Are you driving to work or working to drive? Low cost leaflets by **Smart Moves** and **Semlyen** called 'What are my Car Costs?' and 'Add up your Car Costs' are available from **Smart Moves**. There are also disk versions for Mac or PC which do the mathematics for you.

CAR COSTS EXAMPLE

A new petrol car bought for £13,000, with a 1101-1400cc engine driven 10,000 miles per year costs around £4,200 per year (1999).[1]

1 Purchase Costs

A	My car, with all its extras excluding interest cost	A	£13,000
C	I could sell my car for (phone a dealer or look in your local paper's car ads section then deduct 10%)	C	£11,430
D	I have owned my car for (months)	D	12
L	Each month my car is worth less (A - C) divided by D	L	£131
M	If I had not bought a car, I would have earned interest of a net 3% per year (after tax and inflation) A x 0.03	M	£390
N	My annual purchase cost = L X 12 + M	N	£1,960

2 Other Fixed Costs

A	Car tax	A	£155
B	Breakdown membership	B	£74
C	Car insurance	C	£375
D	MOT test fee (excluding repairs needed to pass!)	D	£29
E	Garage costs (e.g. rent and council tax)	E	-
F	Parking permit costs (home and/or workplace)	F	£40
G	Total fixed costs = A + B + C + D + E + F	G	£673

3 Running Costs

A Miles per year the car is driven	A	10,000
B Miles per litre of fuel on average (miles per gallon divided by 4.546 = miles per litre)	B	7.70
C I buy = A divided by B litres of fuel per year	C	1299
D Fuel costs per litre (in pence)[2]	D	84p
E Total cost of my fuel = (C x D) / 100	E	£1090
F 12 months' parts, servicing costs, repairs and oil	F	£480
G In a year, parking, tolls and car washes cost	G	£30
H Driving fines (parking, speeding etc.)	H	£0
J Variable costs before depreciation = E + F + G + H	3J	£1,600
K True variable running cost with mileage-related depreciation of 3.5p[3] a mile (1998) = J + (0.035 x A)	K	£1,950

4 Ownership costs

A Total purchase costs: 1N	A	£1,960
B Minus mileage-based depreciation of 3.5p a mile (1998) = 3.5p x 3A divided by 100	B	£350
C Total other fixed costs: 2G	C	£673
D Total ownership costs = (A - B) + C	D	£2,283

5 Total annual costs

A Total annual cost of my car = 1N + 2G + 3J	A	£4,233
B Total weekly cost of my car = A / 52.18	B	£81

6 Income Required

A My top rate of income tax is	A	23%
B I pay National Insurance Contributions at	B	10%
C My marginal tax rate is = A + B (33% is standard 1999).	C	33%

D The total yearly income to run my car
   = 5A divided by (100 - C) x 100   D   £6,320
E A weekly income = D divided by 52.18   E   £121

7 Hours Worked

A My hourly wage is   A   12
B I work 6E divided by A hours a week to pay for my car
                                            B   10

These figures are shown in a graph on page 95.

GIVING UP YOUR CAR

Being car-free is a positive move. If a vehicle is needed occasionally, hire a taxi, new or car club vehicle for much less, or share someone else's car. There is a safety net!

Planning is vital to success. Look at your lifestyle. Is it feasible to give the car up? Look at alternative ways to do things, including work, shopping, leisure and escort journeys. Do you have everything else required at your fingertips? An umbrella, waterproofs, alternative travel plans, good shoes, taxi numbers, public transport details, the relevant maps and car hire details etc.? Sort them out or fill details in this book as soon as possible. Of course, cycling is highly recommended as it is cheap, healthy and door-to-door. Do you each have a bike and, ideally, one folder and bike trailer? Are they well maintained?

Preparing to Sell

• Discuss being car-free with co-owners/your household.

• Localise and agree journey plans for all frequent journeys – see pages 74-75 and 152-155.

• Add up your car costs compared to the alternatives – see pages 90-101.

• Collect all relevant public transport timetables and ticket numbers.

• Prepare for not having a car, by trying alternatives and restructuring your life.

SUCCESS STORIES

Retired physics teacher **Rosemary Thornton** from York gave her van to her mechanic when costly servicing, insurance and tax were due. Environmental, safety and quality of life reasons, particularly for children, were secondary motivators. One day she may share a small, more economical car, but only if it were garaged, as her van had been broken into four times. Rosemary advises others to think carefully about the implications. "Work out the yearly savings to make sure it is worth your while. Have warm gloves when carrying shopping in winter." She reaps savings, is healthier through walking and cycling a little more and has less anxiety about cash flow or car crime. She has bought a folding bike. Trains are definitely better for longer trips as she would not otherwise have driven in bad weather, can meet people and knit whilst travelling. Rosemary is worried about the "prestige" aspect of car adverts, which pander to those whose self-esteem is related to what they drive! Sadly, her cat was run over in Spring 2000.

- Convince all concerned that green alternatives are better and decide on a month that you will sell the car, when it is no longer a necessity.

Selling
- Ring local car dealers for quotes or look in a price guide e.g. Parkers.[1] Most owners/adverts vastly overestimate the actual trading price.

- Clean the car. Cost your time and don't overdo it.

- Drive to two or three dealers and agree the best price locally then walk, cycle, get a bus or taxi home.

- Or, place adverts and show callers the car. Cost your time as this takes longer.

- Or, give it to someone who really needs and wants a car and can afford it.

- Or, ring a dismantler, ideally one that who deals in your brand of car. See **Yellow Pages**.

- Or, ring a scrap yard and agree a deal. First Glasgow buses offered to 'Swap your Banger for a Bus' in June 1998. 500+ old cars were exchanged (and scrapped), each for a £560 annual bus pass.[1] First Glasgow won the 1999 Bus Industry Award for a marketing first.[2] A similar scheme has been proposed for London, covering cars that are F-reg or older. If the merchants have to collect the car, they are unlikely to pay you anything. If the car is not drivable, they may not want it! 2000 prices included a £1 charge per tyre to dispose of the rubber or £15-£60 paid for scrap.[3]

Staying Car-free

• Avoid places where cars predominate i.e. give up visiting out of town developments in favour of local or city centre alternatives.

• Keep reminding yourself what you are gaining – including savings from mixing transport modes and home delivery of around £1,400[4] a year (1999).

• Use the cash you've saved for treats such as a meal out – not a flight, please!

• Rent out a drive or garage, or convert a garage to a room for a lodger; or cut rent, mortgage or council tax costs by moving somewhere that doesn't have parking.

SUCCESS STORIES

Marketing executive **Tom Haynes** sold his car in 1995. He lives with his fiancée Louisa in West Hampstead; they use public transport to work and get around the capital, and trains or cheap car hire deals for longer trips. "We were taking our first mortgage on a flat and decided our disposable income was better spent on renovation than tied up in a car. From an investment point of view, this decision has paid dividends!" They don't have an unused car parked on a crowded street five days a week, which would be a waste of money and a liability, as car crime is a significant problem. "When we hire, we get a new vehicle exactly suited to what we are using it for – therefore greater flexibility". Tom and Louisa don't go out of town as often, but are more familiar with their neighbourhood and enjoy urban rambling at weekends.

BUYING A SMALLER VEHICLE

Using a small car limits the damage caused by your driving. Only get a smaller car (rather than give it up) if you need to drive regularly and have tried lift and/or car sharing.

What Type of Vehicle?
• Decide what capacity you need. What is the car for? Private cars should satisfy usual, not peak needs. If for town use, a short length is easier to park e.g. a Smart Car, whilst speeds above 30 mph will rarely be used. Lighter, smaller cars are more fuel efficient. Cars not over 1100cc pay £55 p.a. less road tax (1999). Electric or motorised tricycles not over 150cc pay just £15 p.a. road tax (1999).[1]

Preparing to Sell or Part-Exchange
• Remove all non-essential extras to de-personalise the car.

• Consider where to borrow, buy or hire a trailer or bigger vehicle if needed.

• Ask friends/ring for quotes about moving any occasional large loads.

Selling
• Follow the sales advice on page 103. Also look for/ask about a trade-in.

• Time the sale/exchange to fit your transport needs.

A car uses an estimated 20% of its total lifetime energy expenditure before it reaches the showroom: vehicle manufacture is enormously demanding of energy and raw materials. This is a strong argument for keeping old cars on the road, always assuming they are well maintained.[2]

CAR PURCHASE CRITERIA

Demand the facts you really need to decide which car.

• Whole-life costs e.g. price, fuel efficiency, financing (e.g. lease/hire/buy) maintenance, repairs, depreciation and tax. Fleet magazines and major leasing companies provide whole-life costs. Check the assumptions used.

• Cost per 100 miles for urban and long distance use, plus wear and depreciation.

The **DETR** and **Vehicle Certification Agency** produce car fuel consumption figures.

• Insurance and road tax brackets.

• Expected reliability and depreciation.

• Total annual cost of use based on your expected annual mileage. See pages 95-98.

Safety and Environment
• Check the safety class for passengers, pedestrians and cyclists. Choose a highly visible paint colour.

• An environmental rating for the car and the manufacturer.

• A 'clean' fuel (see pages 45-6) or two fuels. Diesel is worst for health.

• Use the **National Society for Clean Air**'s free guide *Clean Cars: How to Choose One* or the **ETA**'s *Car Buyer's Guide*.

Complain about faulty cars/repairs/servicing to the **Retail Motor Industry Federation, Scottish Motor Trade Association** or the **Society of Motor Manufacturers & Traders**.

SUCCESS STORIES

University administrator **Wendy Love** and her partner, from Benton, Newcastle-upon-Tyne, are determined not to be slaves to the car, nor a 'taxi service' for their four children. "We get a little tired of hearing people say "I have no option but the car." Whilst certainly true in Nashville, Tennessee, where they used to live, it is not true for British cities! "Our first consideration in where to live was easy walking distance of public transport". On arrival from the USA, they cut their car use from 10,000 miles to 6,000 and now drive under 3,000 miles a year for environmental reasons. The family commute to work or school by Metro train. Inspired to "think global, act local", they shop nearby. "We buy at small, local shopping centres, at the fruit and vegetable market, and get cheese, meat and breads at department store food halls. We can pick and choose and find things much more easily and quickly." But Wendy is appalled by the condition of pavements and road crossings, and asks whether ANY drivers know they should give way to pedestrians at side junctions. She advises others to brainstorm different ways of doing usual activities. "Once walking has become a habit, it seems inconceivable to go by car and sit in all that traffic." They live on a main road and suffer 24-hour noise and pollution. "City dwellers deserve better and where else to start to change habits than with oneself?" Benefits are that "the kids are so independent, and love getting about without relying on mum or dad! We spend very little on petrol and see a lot more of our friends by using the Metro or walking than we would in a car."

Chapter 7
Making it work

PERSONAL SAFETY

Bus, rail and coach are all safer than the car.[1] Women are more at risk from violence at home and from people they know than from strangers on the streets.[2]

• Plan the trip, telling family or friends your expected arrival time.

• Try to travel mostly in the day.

• Take the destination address and phone number and money for a taxi back-up.

• Late night revellers can leave 10 minutes early to avoid trouble in a taxi queue.

• Consider a personal alarm. Electronic types are more reliable than gas, especially in the cold.[2] Hold it in your hand, not on a cord around the neck.

• Consider carrying a mobile phone, but do not use it whilst driving.

• Note descriptions or number plates of aggressors.

On the streets
• Walk in the middle of the pavement, using a route that is well used by other walkers at the time of travel. Cyclists can seek routes away from traffic.

• If you regularly go walking or cycling, vary your route and time where possible.

• Wear high visibility clothing, or reflective strips on a bag to be seen. If travelling by a vehicle, maintain it and check that you have good lights.

• Keep bags close to your body or under your jacket or coat. Keep house/car keys out of your bag in a pocket and luggage in sight.

• If you think you are being followed, cross the street. If they follow, cross again. Walk fast to the nearest place where there will be other people.

• If walking behind someone alone, cross the street to show that there is no threat.

Public Transport Users
• Wait for public transport, or a lift, in a well-lit and frequently used place.

• Travel with others or sit near the public transport driver. Kids go First is a **GNER** unaccompanied child service. Book in advance for extra supervision from train staff.

• Sit close to an emergency alarm – ideally in a train compartment near the destination exit. Ring the **British Transport Police** to report incidents.

If Attacked
• If someone grabs your bag, police advice is to let it go rather than risk injury.

• If you can't escape, you can use reasonable force in self-defence. Stay calm. Try to surprise or delay your attacker and escape fast in the safest direction. Shouting, screaming or setting off a personal alarm are all sensible distractions. Self defence training is available from local crime prevention officers ☎...........................

WEAK LINKS

Most journeys are a series of stages: like links on a chain they have weak points. Car users face problems with congestion, destination parking and matching lift sharers.

Walkers need walkable routes and safe crossings at junctions. Cyclists, pram or wheelchair users have difficulties with steps, kerbs and steep hills.

Two-wheelers want traffic calmed routes, crossings, right turns and secure parking.

Public transport users need to get to, and leave at, the right stop and for services to connect.

A journey is only as strong as its weakest link – so plan for them.

• Choose the nearest destination as difficulties generally increase with distance.

• Pre-plan the best route, who to travel with, at what time and where to park securely.

• Cyclists, pram or wheelchair users should plan an easy route with low kerbs and gentle gradients. Use lifts or ramps, not steps. Ask your Council for an access map. Electric bikes or mobility aids can help to tackle hills or headwinds.

• Cyclists can plan to go left when possible and use traffic-free or calmed routes.

• Plan how to get to and from public transport stops or get advice e.g. from the driver, station information centre etc.. Or use a taxi, perhaps pre-booking, e.g. **Virgin** CarLink.

• Refer to timetables for connecting services or ask for advice.

Examples of integration initiatives that have worked include **First Great Western**'s bus-rail links guide. **Connex** rail timetables provide information on through ticketing with bus and rail, e.g. Brighton rail users can buy a whole day's bus use for an additional fee of £1.[1]

A folding bicycle helps to overcome integration problems and is the solution the author chooses. Cycle hire has begun to appear at stations such as London Victoria and Waterloo.

WEATHER

Cold, wet, windy or icy weather can be a barrier to cutting your car use. On the other hand, in good weather it feels great to be out in the fresh air, enjoying more sunlight because being with nature lifts our spirits. The seasons are diverse, as are the solutions. Scandinavian countries have the best traffic reduction records in Europe, so the weather is no deterrent for them! Ask the **Met Office** for a forecast.

• Buy the best modern lightweight, thermal, waterproof and protective clothing (e.g. boots, socks, gloves, hat and scarf) that you can afford. Wear layers. You may also need a foldable umbrella and waterproof bag.

• Perhaps decide not to travel, go shorter distances or allow for longer travel times.

• If cold when walking or cycling, speed up to raise your body heat with activity, or take a flask or stop at a cafe and warm up with a hot drink or snack.

• Undercover or indoor routes may be available i.e. covered walkways.

• Carry or keep spare clothes (e.g. trousers, socks) at work to change into or smarten up with if needed.

• Time the trip to arrive at a public transport stop with a minimum wait.

• Use public transport stops with the best (ideally indoor) shelters.

• Pick sheltered (e.g. tree-lined) routes rather than exposed routes if it is windy.

• Travelling on ice is dangerous by any mode. Do you need to go? If so, go slow.

• Get a taxi or share a lift instead as these travel modes are door-to-door.

CARRYING LOADS

Stop loads from being a barrier to limiting your car use. There are solutions.

• Only take or buy as much as you need and can carry. Persuade others to carry their own loads. e.g. teachers could ask pupils to bring their own kit where possible.

• Use local shops and facilities.

• If walking or cycling, take a rest when you feel tired.

• Ask others to help, or take loads for you. Offer to pay them.

• Leave belongings in station left luggage lockers or other storage place. Ask for storage. Storage at work/school does help limit loads carried home and can be used for cycling gear. Clubs could keep kit in a locked cupboard at their venue.

• Hire a taxi, car or van occasionally and do all your bulky jobs in one go.

• Ask for kerbside recycling or begin recycling with neighbours.

Equipment
• Buy collapsible, lightweight and durable equipment for easy carriage and storage e.g. a collapsible bag, suitcase, folding bike, trailer, trolley, small tent etc..

• Consider duplicate kit, e.g. two copies of a heavy book or garden tools. A friend keeps a spare amplifier at the restaurant where he usually entertains.

• Spread weight evenly e.g. use a strong rucksack with back padding and padded shoulder straps for comfort.

• Push or pull a trailer, trolley, handcart, pram, push chair, case on wheels etc..

• Panniers, baskets and trailers are cheaper than taxis for carrying loads on a bike. Alternatively, push a bike with loads in a basket, panniers or both handlebars.

• The best school escorted walking schemes have a push trolley for school bags.

Low-Cost Load-Carrying Devices[1] is a manual about back-frames, hand carts etc.

Shopping
• When looking for a special product, phone first to check it is in stock or order it.

• Ask about delivery of bulky/heavy items and be prepared to wait a few days. Ask about specific delivery times or ask neighbours to accept goods for you.

• Use home delivery, or shop as usual, then have goods delivered – see pages 23 and 85-87.

• Pay a deposit and pick up goods when convenient.

• Suggest to your Council that your town centre could have a shopping trolley scheme, linked to a home delivery service for products from different outlets.

Public Transport
• On public transport, keep valuables and luggage in sight. Report any crime to the **British Transport Police**.

Cycling
A combination of medium-sized rear panniers, medium size panniers mounted low on the front and a small handlebar bag is the best way to carry loads on a bike. The heavier the object, the lower it should be placed for a low centre of gravity.[2] Some front baskets detach to become bags e.g. **Brompton**. Or use a tricycle or bike trailer.

• Trailers are from **Cycle Heaven**, **York Cycle Works**, **Living Lightly**, **Bike Hod** and **Orbit** among others. **Cycles Maximus** specialise in load-carrying tricycles.

• Buy a dog trailer, e.g. from **Two Plus Two**.

• Ring advertisers in magazines like *Bycycle*, *A to B*, or *On Your Bike* or see *Encycleopedia*.

Many trailers detach from bikes to become trolleys. Consider using child-carrying trailers for other loads. See page 117.

CARRYING CHILDREN

With kids in a car, how often is the journey itself fun? The alternatives give a deeper quality experience – which is good for young, inquiring minds. Driving might seem convenient, but are your assumptions of ease, speed and safety real? In-car air is the most polluted and kids often prefer to walk or cycle. Cycling is quick for short distances.

• Traditional ways of moving children are by pram, push chair or carrying sling. Alternatively, take public transport and keep the kids amused whilst someone else does the driving.

SUCCESS STORIES

DH last owned an 'inherited' car for six months in 1996. He has cut his driving steadily from around 5,000 miles p.a., because "it was a pain to use a car, and I had better things to do." Now he rarely hires, and has adapted his lifestyle to do almost all he needs on a bike, bus or train. "Most times we hire, with a game plan to keep wheels rolling on a series of pick-ups and drops, and minimal idle time. Using the same hire firm gets us some very generous deals." He advises others to plan, "and then realise how much money, time and other resources you save – but most of all be thorough in the understanding of your other possible travel modes." Since selling his car, DH has raised his quality of life, saved "loads of dosh" and reduced his blood pressure and pulse rate. His weight is consistent, he is almost always knows how long a journey across town will take, and can shop in under an hour at any time of day.

Cycling
• Use a child seat or trailer such as the Burley range e.g. from the **UK Trailer Co** or a BycycleR Evolution from **Valley Cycles**. Some convert into prams e.g. the Transit Delux from **Trek**. **Cycle Heaven** and **York Cycle Works** offer mail order.

• Alternatively, get the kids to pedal too on a trailer bike, e.g. Mongoose Alleycat from **Hot Wheels**, the Tagga from **Taylor Cycles**, **Tug** or **Two Plus Two**. The Linkit System from **Astell Leisure Products** is a connector bike system.

• Tricycles or tandems are good for family cycling. The **CTC**, *On Your Bike* and *Encycleopedia* are sources of advice Also see school journeys on pages 80-84. Many trailers can also be used for carrying pets or other loads. See page 115.

SETTING TARGETS

Transport is not usually an end but a means. Deep down, everyone wants good relationships and to feel peace, love and joy. These emotions are promoted by positive thinking or supportive people, not by objects, such as a car.

Huge advertising budgets are lavished on persuading people to buy cars. Ford spent £72 million and Vauxhall £58 million between January and September 1997.[1]

However, contrary to the advertisements, cars don't really bring ultimate freedom. Many benefits disappear in a traffic choked street! So, instead of focusing on movement, see travel as giving access to exchange e.g. goods and services, health and social needs. Near exchanges are better as sustainable, high quality living hinges on localising. Given that exchange decreases with excessive movement, how much mobility do you need or really want? Set your targets.

1. My current car use is .... miles a year.
I aim to cut it by .... % in 6 months, by .... % in a year
to under .... miles p.a.

2. I am usually in a car for .... hours a week.
I aim to be in a car for a maximum of .... hours a week by .... (date).

3. I am usually in a car on .... days a week.
I aim to use a car on only .... days a week by .... (date).

4. My car's engine capacity is ....cc. I aim to get a car with a smaller engine capacity of ....cc and better fuel efficiency by .... (date).

5. My household has ...... car(s). I aim reduce it to ...... car(s) by .... (date).

SUCCESS STORIES

Bob Lavers from York uses his small car 20% less (from 10,000 miles p.a. to 8,000) because of the general unpleasantness and costs of driving. It is also a consequence of retiring from full-time work and hence less need to drive. Bob's 1.3 litre car does 50 mpg on lead-free petrol, but it's cheaper and less trouble to cycle locally, and he benefits from the exercise. Bob advises others to "walk, share cars, take the bus or train – but above all cycle".

SUCCESS

Make an affirmation by saying "I will succeed in driving less." Do it with total intention, i.e. with the desire for, belief in, and acceptance of, less personal car use. You really can make a difference! Because actions follow thought, it is vital to refresh and energise this positive idea in your mind often. Then it will happen and you will become a role model for others.

Practise consciously cutting your car use whilst also not being attached to the results. Then you will not be discouraged if your resolve, or your abilities, fluctuate. Know that with correct practice at cutting your car use you will definitely succeed. Good luck!

Cars are romanticised objects, and dependency on them is a kind of addiction. In this case, habits can change without total withdrawal from the drug. The aim is not to overindulge! Become more creative and self-assured instead. If a private car is not available, you could use a taxi, lift share, car share, car pool, or hire a car when needed.

Check you progress in six and twelve months time. Reward yourself (e.g. with a meal in a local restaurant) when you reach one of your targets. Then set a new goal. Keep reminding yourself of what you are gaining and rejoice in your success at greener travel!

SUCCESS STORIES

Medical research scientist **Jim Young** from Swansea and his wife Susan sold their car for financial reasons, when both their sons were doing PhDs. "We now firmly believe that the only way to change one's use of the car is to give it up altogether". Both walk to work, and would not now consider car ownership unless forced by ill health, as they hate driving. "We always know that we can get a taxi in an emergency." Jim cites two important factors that will enable you to adapt successfully. "Firstly, a gradual increase in fitness, as one develops the muscles for walking". In their case this took about six months, and at ages 51 and 48 they feel fitter than they did three years ago. The second factor is attitude. "We no longer rush to as many locations as we used to. We treat our excursion as a day out which happens to include the location to perform a particular task, such as shopping or work". They use green corridors (e.g. parks) and are more relaxed about imagined time constraints. Adapting to the weather necessitated buying waterproof jackets, leggings and shoes. Motorists often ask them how they cope with being limited to the walkable locality. "On the contrary, we feel that every motorist is trapped by their car dependence – financially, stress due to traffic conges- tion, inflexibility of routes, rushing, drooling over the lat- est car specifications, unfitness, polluted in-car air, an inability to converse in depth with a companion or to stop when it suits the conversation and, finally, the inconvenience of car servicing."

SUCCESS STORIES

Birmingham lecturer **SJHB** cut his driving by 90% in a year by switching to cycling, walking and public transport. He has flexible hours and lives near the city centre. His children go (or went) to schools in walking distance. He only drives to the dump or to his Gloucestershire cottage – which is not on a rail route. "The age of motor touring is over. I experienced a gridlock too many times, and decided driving was not just inconvenient but no fun. I am increasing unfriendly toward 'car-culture' and the assumption that cars are linked positively to a better economy – in fact the reverse is true." He knows many road victims and as a child used to dread hearing that his parents were in an accident. The final motivator was a Sports Lab research project. "On cycling 26 miles for six weeks my fitness improved so markedly that I was convinced cycling was sensible and enjoyable". Freedom, improved physical and mental health and meeting more people are benefits of less car use. He likes being outside and has more city knowledge – "its canals, abandoned railway tracks, routes across parks and beside streams; all things I hardly noticed from my car." SJHB works with Local Authorities and his folding bike allows combinations such as trains, buses, trams, taxis and private cars. "I save my clients sometimes over £25 a day and don't worry about car theft or vandalism." His family still have two cars. "As I prepare finally to abandon my car, I shall rely even more on car-share and taxis."

References

Introduction
1. **DETR**. *Focus on personal travel*. 1998. National Travel Survey 1995/7:1052 journeys per GB resident per year = 2.9 journeys per resident per day.
2. **Transport 2000**. Myths and facts: transport trends and transport policies. 1994 Jun.
3. **Inland Revenue**. Using your own car for work. 1999 Jun (IR125). Ratios of variable to full cost allowances = 55-61% depending on engine capacity. Average = 57%.
4. City of Edinburgh Council. Edinburgh transport factsheet 5. 2000 Feb
5. **DETR**. *Breaking the logjam: the Government's consultation paper on fighting traffic congestion and pollution through road user and workplace charging*.1998; Dec 2.2.
6. **DETR**. Focus on personal travel. 1999. National Travel Survey 1996/8.
7. Devon County Council. **TravelWise** advert. 1998 Sept.

The Vision
1. Stradling SG, Meadows ML & Beatty S. *Factors affecting car use choices*. Transport Research Institute, Napier University 1999 Dec. 33% of 791 drivers would like to use the car less.
2. **RAC** *Report on Motoring 2000*. 39% of 1,563 drivers would use their cars less if public transport were better.

Chapter 1 WHY CUT YOUR CAR USE?

Save Money
1. The **Stationery Office**. *Family spending: a report on the 1996-7 Family Expenditure Survey*.
2. **Smart Moves, Semlyen** A. What are my car costs? 1999 Apr.
3. **DETR**. *Focus on personal travel*. 1998. National Travel Survey 1995/7 average miles per year travelled per GB resident = 6728.
4. **Inland Revenue**. Using your own car for work. 1999 Jun (IR125).
5. **DETR** Driver & Vehicle Licensing Agency. Rates of Vehicle Excise Duty. 1999;Jun: V149.

Be Healthy
1. **ETA**. Road user exposure to air pollution. 1999 Nov.
2. Martin Dr S. Team leader on coronary heart disease prevention, **Department of Health**. Speech at Safe Routes to Schools Conference. York 1999 30 Jun.
3. **Transport 2000** Trust. *A safer journey to school: A guide to school travel plans for parents, teachers and Governors*. London 1999. Free from **DfEE**.

Be Green
1. Vauxhall Motors Ltd. Alternative fuels .. have you considered the options? Luton undated V12634.
2. **Friends of the Earth Trust**. *Road transport & air pollution*. 1999 Mar.
3. **DETR**. *The environment in your pocket*. 1998 Sep.
4. **DETR**. National road traffic forecast. 1997.
5. **DETR**. *Focus on personal travel*. 1998.
6. Whitefield P. The ideal home. *Permaculture* 1998;19:5.
7. Scottish **Green Party**. Air transport policy (draft). 2000 Feb.

Manage Time
1. **DETR**. *Focus on personal travel*. 1998. National Travel Survey 1995/7: GB residents spend 355 hours per year travelling.
2. Hillman M. *Curbing shorter car journeys: prioritising the alternatives*. **Friends of the Earth Trust** 1998 Mar.

Chapter 2   TRAVELLING LESS

Stay Still
1.TRL. *Monitoring and evaluation of a teleworking trial in Hampshire*. Crowthorne 1999: 4/4.

Localise
1. *Car Busters*. Towards car-free cities 1. Proceedings, Lyon 1997.
2. Newman P, Kenworthy J. The forgotten history of automobile development. *Car Busters* 1999 Autumn;10-1.

The Psychology of Changing Travel Modes

1. Swedish National Road and Transport Research Institute. *Analysis and development of new insight into substitution of short car trips by cycling and walking - ADONIS*. Behavioural factors affecting modal choice 1998.

Timetables

1. Brighton & Hove Metro Bus Times free guide. 1998 winter.

Chapter 3   MAKING BETTER USE OF THE CAR

Slow Down

1. Rural roads: the hard facts [editorial]. *Slow Down!* **Slower Speeds Initiative** 1999 Autumn.
2. **DETR**. Speed review: emerging issues and some questions. 1999 Autumn.
3. **DETR**. *Vehicle Speeds Great Britain*. 1998.
4. Spence K, Road Safety Officer, City of York Council. Speech at Slower Speeds Meeting. York 1998 Oct.
5. Pressure grows for lower speeds [editorial]. *Slow Down!* **Slower Speeds Initiative** 1999 Spring.
6. TRL. Taylor, M, Lynam D, Baruya A. *The effects of driver's speed on the frequency of road accidents*. TRL Report 421.
7. The Australian experience [editorial]. *Slow Down!* **Slower Speeds Initiative** 1999 Spring.

Road Safety

1. **DETR**/The **Stationery Office**. *Road Accidents Great Britain*. 1999. 1998 all road users: fatal 3,581; injured 335,033.
2. **DETR**/The **Stationery Office**. *The Highway Code*. 1999.
3. Carlo D. Slow down. *Transport Retort*, **Transport 2000** 1998 Apr;12-3.
4. **DETR**.1998 Valuation of the benefits of prevention of road accidents and casualties. Highways Economics Note 1.
5 OECD Economic importance of accidents. <www.oecd.org/news/events/release/nw99-90a.htm>.
6. The **Slower Speeds Initiative**. Policy briefing 1999.1.
7. The Portman Group. Drinking & driving: what every driver needs to know. undated 1993 figures.
8. Lincolnshire County Council. Avoiding drugs & driving accidents. undated.

9. Douglas N. Sleep Foundation. National Radio 4 News 1999 22 Mar.
10. **DETR**. Mobile phones and driving. London 1998 T/INF/451.
11. **RoSPA**. New research links mobile phones with road accidents. <www.rospa.co.uk/pr16.htm>. 1999. 25 Feb
12. Clarke C. Road Safety Improvement Bill proposals. York 1999.

Traffic Flow
1. **DETR**. *Breaking the logjam: the Government's consultation paper on fighting traffic congestion and pollution through road user and workplace charging.*1998; Dec 2.2.
2. Davis D. Think about it. *Going Green*, **ETA** 1999;33.
3. Department of Transport (DETR). Think before you travel. 1997 Jan T/INF 431.

Parking
1. **DETR**. National Travel Statistics 1999. Reported on BBC online 1999 Oct 14.
2. Clarke C. Road Safety Improvement Bill proposals. York 1999.
3. **DETR**. *Preparing your organisation for transport in the future: the benefits of Green Transport Plans.* 1999.

Energy Efficiency
1. **DETR**. *The environment in your pocket.* 1998.
2. **DETR**. *Running a greener vehicle.* 1998.
3. **Energy Efficiency Best Practice Programme**. *That's an idea.* 1998 Sep.
4. **ETA**. Petrol, diesel or something else? <www.eta.co.uk/fact/diesel.htm>.
5. Vauxhall Motors Ltd. Alternative fuels . . . have you considered the options? Luton, undated.
6. *Permaculture* magazine (ad) 1999;19:2

Air Quality
1. **Friends of the Earth** Trust. *Road transport & air pollution.* 1999 Mar.
2. **DETR**. Winter smog, summer smog. 1998 Jul 98/EPO484.
3. **DETR**. Air pollution inquiry line. 1999 Dec.
4. **DETR**. Air pollution - what it means for your health. 1998 Jul 98/EPO37.
5. **ETA**. Road user exposure to air pollution. 1997 Nov.
6. Maddison D, Pearce D, Johansson O, Cathrop E, Litman T, Verhoef El. *Blue Print 4: The true cost of road transport.* Earthscan 1996, Box 4.11.

7. **DETR**. Running a greener vehicle. 1998 Jul 97/EO484.
8. Collinge W. *Subtle energy*. Thorsons 1998;90. Reporting research by the Smithsonian Institution.

Cleaner Fuels
1. **ETA**. Petrol, diesel or something else? <www.eta.co.uk/fact/diesel.htm>.
2. **Energy Efficiency Best Practice Programme**. *That's an idea*. 1998 Sep.
3. Vauxhall Motors Ltd. Alternative fuels . . . have you considered the options? Luton, undated.
4. Plowden S. Vehicle design. *Going Green*, **ETA** 1999 Spring.
5. **DETR**. Still running on leaded petrol? 1999 Jul.

Quiet
1. **DETR**. *The environment in your pocket*. 1998 Sep.
2. Dept of the Environment (now **DETR**). Bothered by noise? There's no need to suffer. 1997 Mar.
3. **Transport 2000** et al. *Getting out of neutral: how the Government can move forward on transport policy*. 1999 Sep.

Chapter 4   WHAT ARE THE ALTERNATIVES?

Walking
1. **ETA**. Road user exposure to air pollution. 1997 Nov.
2. North Yorkshire Specialist Health Promotion Service. Why walk? undated.
3. **Semlyen** A. The price ain't right. *Bycycle* 1999;6:48-9. Walking is 7p a mile (£125 extra in shoes & clothing over 5 years for 1 mile more daily).

Lift Share / Ride Share
1. **Inland Revenue**. Using your own car for work. 1999 Jun (IR125).
2. **Smart Moves, Semlyen** A. What are my car costs? 1999 Apr.
3. Brown G. Budget speech 1999.
<www.hm-Treasury.gov.uk/budget99/speech.htm>.

Taxis
1. City of York Council. Hackney carriage fares. 1999 Sep.

Cycling
1. Reid C. Transport of delight. *On Your Bike* 1999;3 Spring.
2. **ETA**. Road user exposure to air pollution. 1997 Nov.
3. Hillman M. *Curbing shorter car journeys: prioritising the alternatives.* **Friends of the Earth** Trust 1998.
4. Illich I. Energy and Equity, in *Toward a history of needs*. Pantheon 1978.
5. UKK Institute. Patient education and counselling 1998;33. Tampere trial Finland. Reported in *On Your Bike* 1999 Spring.
6. Boyd, Hillman, Nevill, Pearce, Tuxworth. Health related effects of regular cycling on a sample of previous non-exercisers. *Bike for Your Life*. **CTC**.
7. **DETR**. *Preparing your organisation for transport in the future: the benefits of Green Transport Plans*.1999 Jun.
8. **CTC**. Pavement cycling fines [editorial]. *Cycle Touring and Campaigning* 1999 Oct/Nov.
9. Curnow B. Road rules OK? *Current Affairs Bulletin* 1998 Apr/May.
10. Brown G. Budget speech 1999.
<www.hm-Treasury.gov.uk/budget99/speech.htm>.

Buying A 'Green' Bicycle
1. Rosen P. Pedals and Principles. *Bike Culture Quarterly* 1995;5:60-1.

Cycle Security
1. **DETR**. *Cycling in Great Britain*. 1996 Aug.
2. Newton E. Cycle security. *Bycycle* 1998;3:38-42.

Electric Bicycles
1. Tomlinson S. Charge of the bike brigade. *On Your Bike* 1998;3 Summer:68-72.

Bus / Park & Ride
1. Department of Transport (now DETR). Think before you travel. 1997 Jan.
2. **ETA**. Road user exposure to air pollution. 1997 Nov.
3. Potter S. *Vital Travel Statistics* 1997. Casualties per km index compared to a car = 1, bus = 0.5. Casualties per billion journeys: bus = 0.4 1994 GB.
4. James A. Exploding myths about the cost of car transport. *World Transport Policy & Practice* 1998;4(4):10.

Train / Underground
1. Potter S. *Vital travel statistics* 1997. Casualties per billion km & per billion journeys 1994 GB.
2. **Inland Revenue**. Using your own car for work. 1999 Jun (IR125).
3. Great Britain Rail Passenger Timetable 1999.
4. Great Britain Train Fares Manual (£13.50 1999).

Coach
1. **DETR**. *Transport statistics Great Britain*. 1998. Killed or seriously injured per passenger km 1994: Bus/coach = 20; Car/taxi = 40.
2. **National Express** <www.nationalexpress.co.uk> 2000 Jan.

Hire Cars
1. York's standard taxi rates (Sept 1999) were: first 535 metres or 2 mins 37 seconds £1.70. Each succeeding 153 metres or 45 seconds 10p· Every mile beyond the first costs £1.05p. Versus a weekday hire of £27 plus petrol @ 11p p/mile. 30 miles = £32.85 taxi fare. Petrol for hire car = £3.30. Car hire cost = £30.30 for 30 miles.
2. National Car Hire, York. Small car £27 day, £55.35 weekend, £170 week. 10% discount after three hires. 1999 Mar.
3. Compares fixed ownership costs of £1,320 to above hire rates with 10% discount after three hires.
4. **Smart Moves, Semlyen** A. What are my car costs? 1999 Apr.

Car Share
1. **Smart Moves, Semlyen** A. What are my car costs? 1999 Apr.

Car Pools
1. **Transport 2000**. *Changing journeys to work*. 1997.
2. **Inland Revenue**. Using your own car for work. 1999 Jun (IR125).
3. Robinson K. *Companies and cars: the way forward*. **DETR, Ashden Trust**, London First. undated circa 1997.

Car Clubs
1. Mobility CarSharing. *Mobility Journal*. 1999 Apr.
2. Haydon R. Street fleets. *Transport Retort*, **Transport 2000** 1998;Apr:10-11.

Moped, Scooter, Motorbike & Trike
1. **DETR** *Road Accidents Great Britain*, 1998.
2. **DETR**/Driver & Vehicle Licensing Agency. Rates of Vehicle Excise Duty. 1999 Jun.

Other Alternatives
1. Anti Gravity Shop, York. 1999 Jun.
2. **DETR/The Stationery Office**. *The Highway Code* 1999. Rules 215 & 245.
3. Engel C. PermaPonies. *Permaculture* 1999;20:33-35.
4. Rees L. *Understanding your pony*. Stanley Paul.

Chapter 5   WHY ARE YOU TRAVELLING?

Commuter Journeys
1. **DETR**. *Preparing your organisation for transport in the future*. The benefits of green transport plans. 1999 Jun.
2. Survey on Commuting [editorial]. *The Economist* 1998 Sep 5.
3. **DETR**. *Focus on personal travel*. 1998. National Travel Survey 1995/7.

Business Journeys
1. **Semlyen** A. The price ain't right. *Bycycle* 1999;6:48-9.

School Journeys
1. **Smart Moves, Semlyen** A. What are my car costs? 1999 Apr.
2. Department of Transport (now **DETR**). National Travel Survey 1994-6.
3. **Sustrans**. *Safety on the streets for children*. 1996 Nov.
4. **Sustrans**. National Cycle Network Catalogue. 1999.

Food Shopping Journeys
1. Cairns S. Menu for change. *Transport Retort*. **Transport 2000** 1997 Sept/Oct.

Chapter 6   HOW MUCH WILL YOU SAVE?

Cycle Costs
1. **CTC**. *What price cycling?* 1993.
2. **Semlyen** A. The price ain't right. *Bycycle* 1999;6:48-9.

Mixed Transport Modes and Costs
1. **Smart Moves, Semlyen** A. What are my car costs? 1999.
2. Minster Bus Pass. York 1999 Oct.
3. **Semlyen** A. The price ain't right. *Bycycle* 1999;6:48-9.
4. City of York Council. Hackney carriage fares. York 1999 Sep.
5. National Car Hire. Small car £27 per day, £55.35 weekend, £170 per week with 10% discount after three hires. York 1999 Mar.
6. The **AA**. Fuel price survey. 1999 Dec 4. National average price for super unleaded petrol.

Car Allowances and Costs
1. **Inland Revenue**. Using your own car for work. 1999 Jun (IR125).

Car Purchase Costs
1. **Smart Moves, Semlyen** A. What are my car costs? 1999 Apr.

Car Costs Example
1. The **AA**. Motoring costs 1999. New petrol car costs 1101-1400 cc (interest & parking costs added).
2. The **AA**. Fuel price survey. 1999 Dec 4. National average price for super unleaded petrol.
3. James A. Exploding myths about the cost of car transport. *World Transport Policy & Practice* 1998;4(4):10.

Giving Up Your Car
1. EMAP National Publications. *Parker's car price guide.* London 1998.
2. The Bus Industry Awards Ltd. *The Bus Industry Awards 1999 Souvenir Brochure*:11.
3. Elvington Autospares, York. Advice from a car dismantler. 2000 Mar.
4. Haydon R. Street fleets. *Transport Retort.* **Transport 2000** 1998;Apr:10-1.
5. Scottish **Green Party**. Air transport policy (draft). 2000 Feb.

Buying a Smaller Vehicle
1. **DETR**/Driver & Vehicle Licensing Agency. Rates of Vehicle Excise Duty. 1999 Jun.
2. **ETA**. Petrol, diesel or something else? <www.eta.co.uk/fact/diesel.htm>.

Chapter 7   MAKING IT WORK

Personal Safety
1. Potter S. *Vital travel statistics* 1997. Casualties per billion journeys 1994.
2. Safe on the street? [editorial] *Which?* 1994 Dec.

Weak Links
1. Salveson P. *Getting the best from bus and rail in rural communities: a review of best practice and recommendations for future development.* Transport Research & Information Network. Huddersfield. 1999 Nov.

Carrying Loads
1. Dennis R, Smith A. *Low-cost load-carrying devices.*
2. **Bicycle Association**. Cycle bags: how to carry things on a bicycle. 1995.

Setting Targets
1. Baird N. *The estate we're in: who's driving car culture*? Indigo 1998;148.

GENERAL
Baird N. The estate we're in: who's driving car culture? Indigo 1998.
Engwicht D. *Reclaiming our cities and towns: better living with less traffic.* New Society 1993.

Directory
(as at March 2000)

Adshel Centre, 55 Philbeach Gardens, Earls Court, London SW5 9EB ☎020 7591 8900. <www.adshel.com>. Street furniture & cycle loan scheme.
Air Pollution Inquiries ☎0800 556677 (freephone) CEEFAX: 410-417 TELETEXT: 106. <www.environment.detr.gov.uk/airq/aqinfo.htm>. Regional pollution summary, forecast & health info.
Alternative Technology Association (ATA) see **CAT**.
Alternative Vehicle Technology (AVT), Blue Lias House, Station Rd, Hatch Beauchamp, Somerset TA3 6SQ ☎01823 480196 Fax 01823 481116. <rfowler@avt.uk.com>.
Anglia Railways, Ipswich Stn, Burrell Rd, Ipswich IP2 8AW ☎01603 764776 tickets weekdays (08.00-22.00), weekends (08.00-20.00) update ☎01473 693369. Ipswich ☎01473 693469. Colchester ☎01206 571133.
Are you doing your bit? Sitel House, Timothy's Bridge Rd, Stratford upon Avon, Warwickshire CV37 9HY ☎0345 868686 (local). <www.doingyourbit.org.uk>. Free booklet 'Every little bit helps'.
Armitage, Richard, Oxford House, Smithy Fold Rd, Hyde SK14 5QY ☎0161 368 6603. <www.ratransport.co.uk> <info@ratransport.co.uk>. Green transport consultant.
Ashden Trust (The), 9 Red Lion Court, London EC4A 3EF ☎020 7410 0330. Transport research publications & grants.
Association of Cycle Traders (ACT), 31A High St, Tunbridge Wells, Kent TN1 1XN ☎01892 526081 Fax 01892 544278. <www.cyclesource.co.uk> Retailer organisation.
Association for Commuter Transport ☎020 8741 1516 Fax 020 8741 5993. <mail@act-uk.com>. Encourage green transport plans.
Astell Leisure Products, Back Lane, Pilsley, Chesterfield, Derbyshire S45 8HJ ☎01773 872407. Connectors to link child to adult bikes.
A to B **Magazine**, 19 West Pk, Castle Cary, Somerset BA7 7DB ☎01963 351649 Fax 0870 052 0810. <www.a2bmagazine.demon.co.uk> <post@a2bmagazine.demon.co.uk>. Cutting car use, alternative transport, folding & electric bikes, trailer & public transport magazine. 6 issues £10 UK (2000).

Automobile Association (AA) ☎0990 500 600 (national).
<www.theaa.co.uk> Motoring organisation. Technical Info, Lister Point,
Sherrington Way, Basingstoke, Hampshire RG22 4DQ ☎01256 491493
Fax 01256 491490. Car costs info. National Byway maps ☎01425 650166.
A–Z Maps, Fairfield Rd, Borough Green, Sevenoaks, Kent TN15 8PP
☎01732 781000. <www.a-zmaps.co.uk> <digital@a-zmaps.co.uk>.
Bartholomew's ☎0141 306 3100. <www.bartholomewmaps.com>.
Maps, mail order sales.
Bicycle Association (BA), Starley House, Eaton Rd, Coventry CV1 2FH
☎024 7655 3838 Fax 024 7622 8366. <info@bicycle-association.org.uk>
Represents cycle industry. A4 SAE (40p stamp) for free funding cycle
schemes booklet.
Bicycling Books, 164 Eign St, Hereford HR4 0AP ☎01432 340 666.
<www.bikebook.demon.co.uk> <sales@bikebook.demon.co.uk>. Largest
cycling book stock in Europe. Free catalogue.
Bike Hod, Two Plus Two, 31 Western Road, Lewes, E Sussex BN7 1RL
☎01273 480479. <info@twoplustwo.uk.com>. Cycle trailer & trailer bike
specialists.
Bikepark, 11-13 Macklin Street, Covent Gdn, London WC2 5NH
☎020 7430 0083. The Courtyard, 250 King's Rd, London, SW3 6NT
☎020 7565 0777. <www.bikepark.co.uk> <office@bikepark.co.uk>.
Secure bike parking, sales and repairs.
Birdy, Riese und Muller, Erbacher Str. 123 D-64287 Darmstadt,
Germany ☎+49 6151 424034 Fax +49 6151 424036. <www.r-m.de>
<team@r-m.de>. Folding bikes.
Brake, PO Box 548, Huddersfield HD1 2XZ. ☎01484 559909
Fax 01484 559983. <brake@brake.org.uk>. Charity for safety & victims.
Organise Road Safety week (July) and Road Risk Forum.
British Allergy Foundation, 30 Bellegrove Rd, Welling, Kent DA16 3PY
☎020 8303 8583. <www.allergyfoundation.com>
<allergybaf@compuserve.com>. £10 p.a. (2000).
British Cycling Federation (BCF), National Cycling Centre, 1 Stuart
St, Manchester M11 4DQ ☎0161 230 2301 Fax 0161 2310591.
<www.bcf.uk.com> <info@bcf.uk.com>. Cycle racing & encouraging
young people to cycle. Network of over 1000 clubs.
British Lung Foundation (BLF), 78 Hatton Gdn, London EC1N 8JR.
Helpline ☎01736 364 365. Office ☎020 7831 5831 Fax 0120 7831
5532. <www.lunguk.org> <blf_user@gpiag-asthma.org>. Organise
Breathe Easy week & local groups. Free leaflet on exercise & the lungs.

British Horse Society, Stoneleigh Deer Pk, Kenilworth, Warks CV8 2XZ ☎01926 707813. <www.bhs.org.uk> <enquiry@bhs.org.uk> Twenty trailguide books on bridleways & cycleways.

British Human Power Club, 15 Station Rd, Dyce, Aberdeen AB21 7BA ☎01224 772164. <www.bhpc.org.uk>. Promote innovation & utility in fast, comfy Human Powered Vehicles (HPVs) & recumbent cycles. Quarterly magazine/booklet £10 p.a. (2000).

British Schools Cycling Association (BSCA), 21 Bedhampton Rd, North End, Portsmouth, Hampshire PO2 7JX ☎02392 642226 Fax 02392 660187. <www.bsca.org.uk>. Train children for leisure and competition cycling, and adults to cycle with children. £5 p.a. (2000).

British Tourist Authority, Thames Tower, Blacks Road, Hammersmith, London W6 3EL ☎0208 846 9000. <www.visitbritain.com>. English Tourism Council. ☎0208 563 3156 (Titles). <www.englishtourism.org.uk> <fulfilment@englishtourism.org.uk>. 'Where to stay' guides.

British Transport Police ☎0800 405040 (freephone). Policing the railways.

British Waterways, Willow Grange, Church Rd, Watford WD1 3QA ☎01923 201120 <enquiries.hq@british_waterways.co.uk>. Many towpaths offer safe cycling. Permits sometimes needed. National Cycle pack £5.

British Wheel of Yoga, 1 Hamilton Place, Boston Rd, Sleaford, Lincs NG34 7ES ☎01529 306851. <www.bwy.org.uk> <wheelyoga@aol.com>. Charity; A5 SAE for teacher list.

Bromakin Wheelchairs, 12 Prince William Rd, Loughborough, Leics LE11 5GU ☎01509 217569. <www.bromakin.co.uk> <peter@bromakin.co.uk> Hand-powered recumbent tricycles.

Brompton Bicycle Ltd, Kew Bridge Distributions Centre, Lionel Rd, Brentford, Middx TW8 9QR ☎020 8232 8484 Fax 020 8232 8181. <www.bromptonbicycle.co.uk> Folding specialists.

BT Phone Base ☎0800 919199 (freephone).

Bus Appeals Body, c/o National Federation of Bus Users, PO Box 320, Portsmouth, PO5 3SD. Independent bus complaints review.

Bycycle Club, Freepost (SCE 8520) Chippenham, Wilts, SN15 3HD ☎0870 2402128 (national). <www.bycycle.com> Discount club, *Bycycle* magazine – see **Open Road**. Individual £29 p.a, family £39 p.a. with 3rd party insurance (2000).

Byways & Bridleways Trust, PO Box 117, Newcastle upon Tyne NE3 5YT ☎ & Fax 0191 2364086. BBT@highwayman.demon.co.uk Rights of way & countryside access. *Byway & Bridleway* journal.

Cambridge Cycling Campaign, PO Box 204, Cambridge CB4 3FN
☎01223 504095. <www.ccdc.cam.ac.uk/camcycle>
<camcycle@pobox.co.uk>.

Camping & Caravanning Club, Greenfields House, Westwood Way,
Coventry CV4 8JH ☎024 7669 4995 Fax 024 7669 4886.
<www.campingandcaravanningclub.co.uk>. Free sites guide, maps,
weather call. £27.50 p.a. + £4 joining fee (2000).

Car Busters, Krátká 26, 100 00 Praha 10, Czech Republic ☎+420 2 781
08 49 Fax +420 2 781 67 27. <www.carbusters.org>
<carbusters@ecn.cz>. Building & maintaining the international anti-car
movement with a quarterly magazine & resource centre. £2 per issue, £9
p.a. (2000).

Car Free Cities Network, 18 Sq de Meeus, B-1050, Brussels, Belgium
☎+32 2552 0883 Fax +32 2552 0889. <cfc@eurocities.be>. Includes
more than seventy European cities. *Car Free Cities* magazine.

CASCA Compensation & Advice Service for Cycling Accidents ☎0800
542 0196 (freephone). 'No win no fee' cycle lawyers.

Central Trains, PO Box 4323, Birmingham B2 4JB ☎0870 0006060
(national) tickets weekdays (08.00 - 20.00) update ☎0870 0015243.
<www.centraltrains.co.uk>

**Centre for Alternative Technology (CAT) / Alternative
Technology Association (ATA)**, Machynlleth, Powys SY20 9AZ
☎01654 702400 Fax 01654 702782. <www.cat.org.uk>
<cat@catinfo.demon.co.uk>. Charity: visitor centre, publications.
Membership organisation is ATA. *Clean Slate* journal.

CAT Electric ☎ & Fax 01604 864916. Electric bikes.

Chase Organics, Riverdene Business Pk, Molesey Rd, Hersham, Surrey
KT12 4RG ☎01932 253666 Fax 01932 252707.
<chaseorg@aol.com>. Online organic seed catalogue in development.

Chevron Handcycles, Unit 18, Summers Rd, Brunswick Business Pk,
Liverpool L3 4BL ☎0151 707 1146. Lightweight hand-powered cycles &
wheelchair adaptations.

Child Accident Prevention Trust (CAPT), Clerk's Court, 18-20
Farringdon Lane, London EC1R 3AU. Charity to prevent, research & eval-
uate child accidents.

Children's Play Council, 8 Wakley St, London EC1V 7QE ☎020 8743
6016 Fax 020 7278 9512. <www.ncb.org.uk/cpc.htm>
<cpc@ncb.org.uk>. Promote home zones.

Chiltern Railways, Western House, 14 Rickfords Hill, Aylesbury HP20 2RX
☎08705 165165 (national) tickets daily (07.00-20.00) update ☎01494
443497. <www.chilternrailways.co.uk>.
Companion Cycling c/o CVS, 1 Princes Street, Richmond TW9 1ED
☎0961 344 545. Special needs duo-cycling, side by side. £10 p.a. £3 per
hr, Bushy Pk. Middlesex.
Community Car Share Network, The Studio, 32 The Calls, Leeds LS2
7EW ☎0113 234 9299 Fax 0113 242 3687. <www.carshareclubs.org.uk>
<office@carshareclubs.org.uk>. Not-for-profit, limited company support-
ing city car clubs & rural car share schemes.
Community Transport Association, Highbank, Halton St, Hyde,
Cheshire SK14 2NY ☎& Fax 0161 366 6685. Advice, information &
training for organisations running not-for-profit community transport.
Connex, 3 Priory Rd, Tonbridge TN9 2AF ☎0870 6030405 (national)
tickets & update. Trains.
Consortium of Bicycle Retailers Limited (CoBR), Courtyard Loft,
Union St, Newport Pagnell, MK16 8FT ☎01908 613263.
<thecobr@aol.com>. Marketing, CycleList guide.
CPRE (Council for the Protection of Rural England), Warwick
House, 25 Buckingham Palace Rd, London SW1P 0PP ☎020 7976 6433.
<www.greenchannel.com/cpre> <info@cpre.org.uk>. Charity protecting the
countryside, reducing traffic growth. Charter for country lanes & slower
speeds. Publish campaigners guides, leaflets, magazine *Voice*. Individual
£17.50, joint £23, under 25 £12.50, concs £10, family £27.50 (2000).
CTC (Cyclists' Touring Club), 69 Meadrow, Godalming, Surrey GU7
3HS ☎01483 417217 Fax 01483 426994. <www.ctc.org.uk>
<cycling@ctc.org.uk> or <membership@ctc.org.uk>. Working for cycling,
mail order bookshop, National Bike Week & Bike to Work Day. Freewheeler
Cycle insurance available to non-members. Cycle hire list on website.
Cycleaid ☎0808 100 9995. 'No win no fee' cycle lawyers.
Cycle Campaign Network (CCN), 54-57 Allison St, Digbeth,
Birmingham B5 5TH. <cyclecampaignnetwork@bigfoot.com>. Local
group details, conferences & *CCN News*.
CycleCity Guides, Wallbridge Mill, The Retreat, Frome, Somerset BA11
5JU ☎01373 453533. <www.dome.demon.co.uk>
<cyclecity@dome.demon.co.uk>. Urban cycle map publisher.
Cycle Heaven, 2 Bishopthorpe Rd, York YO23 1JJ ☎01904 636578.
<www.cycle-heaven.co.uk> <andy@cycle-heaven.co.uk>. Alternative
transport, 5 cycle trailers for children & luggage.

Cycle Training, 32 Carden Rd, London SE15 3UD ☎020 7564 5990 Fax 020 7732 6639. Cycle Training SE, 17 Howard Road, Brighton BN2 2TP ☎01273 691353.<www.cycletraining.co.uk> <info@cycletraining.co.uk>. All age tuition in safe on-road cycling in Greater London and Sussex. Free to certain groups.

Cycles Maximus, The Bell, 103 Walcot Street, Bath BA1 5BW. ☎01225 319414 Fax 01225 334494. <www.cyclesmaximus.com> <sales@www.cyclesmaximus.com>. Load-carrying cycles.

Cyclists' Public Affairs Group (C-PAG) ☎01483 417 217. Local cycle campaign contacts.

Cyclist's Sourcebook, Mark Allen Publishing Ltd, Jesses Farm, Snow Hill, Dinton, Nr Salisbury, Wilts SP3 5HN ☎0800 137201 (freephone). <leisure@markallengroup.com>. Trade bible.

David & Charles ☎01626 323200. <mail@davidandcharles.co.uk>. Maps.

Department for Education & Employment (DfEE), Publications, PO Box 5050, Annesley, Notts NG15 0DJ ☎0845 602 2260 (local). <www.dfee.gov.uk> <dfee@prologistics.co.uk>.

Department of Environment, Transport & the Regions (DETR), Free Literature, PO Box 236, Wetherby, West Yorks, LS23 7NB ☎0870 1226 236 (national) Fax 0870 1226 237. 3/34 Great Minster House, 76 Marsham St, London SW1P 4DR ☎020 7944 3000. Charging & Local Transport Division ☎020 7944 2478 for free Traffic Advisory Leaflets. Travel Awareness ☎020 7944 4094. Green Transport <www.local-transport.detr.gov.uk/travelplans/index.htm>. The website lists email addresses: e.g. integrated transport is <tdp@detr.gov.uk>.

Department of Health, Richmond House, 79 Whitehall, London SW1A 2NS ☎020 7210 4850 (M-F 10.00-17.00). <www.doh.gov.uk> <dhmail@doh.gsi.gov.uk>. Health advice.

Di Blasi, Sherwood Enterprises,14 Chart House Rd, Ash Vale, Aldershot GU12 5LS ☎01252 329783 Fax 01252 679690. Mopeds & folding bikes.

Directory Enquiries ☎192 <www.bt.co.uk/phonenetuk/>.

Dixon, L. & C. ☎01341 440256. <mawddach@gn.apc.org>. Permaculture training for horse & pony needs. Approved student of the Monty Roberts concept.

Easybike Electric Cycles, Ballantyne Business Centre, Hawkins Rd, Colchester, Essex CO2 8JT ☎01206 792171 Fax 01206 793829. <steve@zone1.co.uk>. Electric bicycles & tricycles mainly for the elderly & less abled. Colour brochure/newsletter.

Edinburgh Car Club, Budget Rent A Car ☎0131 453 5300.

Electric Bikes, EV Select Ltd, 17 Nellfield Rd, Crieff, Perthshire, Scotland PH7 3DU ☎01764 655331 Fax 01764 655441. <www.evselect.co.uk> <info@evselect.co.uk>. Electric bikes & scooters.

Electric Vehicle Co, The Fands, Farnham, Surrey GU10 1PX. ☎0800 980 8789 (freephone) Fax 01252 782972. Electric vehicles.

Electric Vehicle Association, 17 Westmeston Ave, Rottingdean, E Sussex BN2 8AL ☎01273 304064 Fax 01273 390370. <EVA@gwassoc.dircon.co.uk>. *Batteries International* magazine.

Encycleopedia Guide to alternatives in cycling. see **Open Road**

Energy Efficiency Best Practice Programme, ETSU, Harwell, Didcot, Oxfordshire OX11 0RA ☎0800 585794 (freephone). Advice line ☎0541 542 541 (national) Office ☎01235 433302 Fax 01235 432923. <www.energy-efficiency.gov.uk> <envirohelp@etbpphelpline.demon.co.uk>. Guidance for employers on travel plans.

Energy Saving Trust, 21 Dartmouth St, London SW1H 9BP Hotline ☎0845 6021425 (local) <www.est-powershift.org.uk>. Powershift pack on alternative fuelled vehicles.

Entec, Grove House, Whitehorse Street, Baldock, Herts SG7 6QF ☎01462 499599 Fax 01462 499555. Lift share databases.

Environment Agency 24 hr hotline ☎0800 807060 (freephone, open 09.00-17.00 weekdays) to report emergencies. ☎0845 933 3111 office. <www.environment-agency.gov.uk>. Covers England & Wales.

ETA (Environmental Transport Association), 10 Church St, Weybridge KT13 8RS ☎01932 828882 Fax 01932 829015. <www.eta.co.uk> <members@eta.co.uk>. Ethical breakdown & cycling assistance, insurance. Organise Green Transport Week & Car Free Day (Sep). *Going Green* magazine £20 membership. Owned by a charity.

Eurolines UK, 4 Cardiff Road, Luton LU1 1PP ☎0990 143219 (national). <www.eurolines.co.uk> <welcome@eurolines.com.uk>. European coaches.

Eurostar declined a free listing!

First Great Eastern, North Station, Colchester, Essex CO1 1XD ☎08459 505000 (local) tickets (06.30-22.30) update ☎020 7247 5488. <customer-services.ger@ems.rail.co.uk>. Trains.

First Great Western Trains, Milford House, 1 Milford Street, Swindon SN1 1HL ☎08457 48 49 50 for rail enquiries, ☎08457 000 125 for tickets. <www.greatwesterntrains.co.uk> <www. traindirect.co.uk>.

First North Western ☎0870 6066007 (national) tickets, update ☎08457 484950 (local). <http://.nwt.rail.co.uk> <customer.relations.nwt@ems.rail.co.uk>. Trains.

Forestry Commission, 231 Corstorphine Rd, Edinburgh EH12 7AT ☎0131 334 0303. Cycle access woodland maps.

Freewheelers <www.freewheelers.co.uk> <freewheelers@freewheelers.co.uk>. Free service linking drivers & passengers for lift share.

Friends of the Earth, 26-28 Underwood St, London N1 7JQ ☎020 7490 1555 Fax 020 7490 0881. <www.foe.co.uk> <info@foe.co.uk>. Charity to protect & improve conditions for life on Earth now and for the future. Publish *Earth Matters* quarterly & other info.

Future Forests ☎01963 350465. <annabel@futureforests.com>. Carbon-audit scheme. Average drivers need to pay for 5 tree plantings p.a. (£20 p.a., 2000) to offset their yearly carbon emissions. .

Gatwick Express, 52 Grosvenor Gardens, London SW1W 0AU ☎0990 301530 (national) tickets & update. <www.gatwickexpress.co.uk>. Trains.

Going for Green, Elizabeth House, The Pier, Wigan WN3 4EX ☎0800 783 7838 (freephone) Fax 01942 824778. <www.gfg.iclnet.co.uk> <ecocal@gfg.co.uk>. Free environmental information.

Great North Eastern Railway (GNER), HQ, Station Rd, York YO1 6HT ☎08457 225 225 (local) tickets & update daily (08.00-22.00). <ww.gner.co.uk>. Allow two working days to post tickets.

Green Books, Foxhole, Dartington, Totnes, Devon TQ9 6EB ☎01803 863260 Fax 01803 863843 <www.greenbooks.co.uk> <sales@greenbooks.co.uk>. Publish a wide range of books on environmental & social issues including organics, eco-building & eco-politics.

Green Party, 1a Waterlow Rd, London N19 5NJ, Freepost Lon 6780, London N19 5BR ☎020 7272 4474 Fax 020 7272 6653. <www.greenparty.org.uk> <office@greenparty.org.uk>.

Greenpeace, Canonbury Villas, London N1 2PN ☎0800 269 065 (freephone) ☎020 7865 8100. <www.greenpeace.org.uk> <info@uk.greenpeace.org>. Non-violent action on climate change & energy.

Health Development Agency, Trevelyan House, 30 Great Peter Street, London SW1P 2HW ☎020 7222 5300 <www.active.org.uk>. Healthy travel advice; also physical activity specialists.

Health Information (NHS) ☎0800 665544 (freephone) weekdays (10.00-17.00). Scotland ☎0800 224488. Free info on weight loss, respiratory problems etc.

Heathrow Express, 4th Fl, Cardinal Point, Newall Rd, Hounslow, Middx TW6 2QS ☎0845 6001515 (local) tickets & update. Trains.

Help the Aged Senior Line ☎0808 800 6565 (freephone) weekdays (09.00-16.00). Welfare rights.

HOP Associates, 55 West St, Comberton, Cambridge CB3 7DS ☎01223 264485 Fax 020 7570 0820. <www.hop.co.uk> or <www.flexibility.co.uk/telecommuting2000> <info@hop.co.uk> Information & communication consultants, mainly to firms. Promote sustainable work & travel behaviour. Publish *Flexibility* & *Telecommuting 2000* (free summary).

Hot Wheels, 1145 Christchurch Road, Bournemouth BH7 6BW ☎01202 424945 Fax 01202 424602. <www.mongoose.com>. Mongoose bicycles and trailers.

Iceland Frozen Foods. Home shopping catalogue ☎0800 3280800 (freephone). Orders ☎0870 2422242 (national) <www.iceland.co.uk>. Delivery within 10 miles of store (3 miles in M25). Free delivery. £4 admin charge to pack your shopping (2000). Internet order trials.

Inland Revenue see Tax Enquiry Centre in local phone book or ☎08459 000444 (local). <www.inlandrevenue.gov.uk>.

Institute of Advanced Motorists (IAM), IAM House, 359-365 Chiswick High Rd, London W4 4HS ☎020 8994 4403 Fax 020 8994 9249. <www.iam.org.uk>. Guidance on safety for drivers.

InTandem ☎01509 843752. Disability cycling.

ISI ☎0345 15 2000 (local). <www.isi.gov.uk> <info@isi.gov.uk>. Business teleworking IT guide.

Island Line, Ryde St Johns Rd Station, Ryde, Isle of Wight PO33 2BA ☎08457 484950 for tickets & update. <www.island-line.co.uk> Trains.

Journey Call, Nelson Rd, Newport, Isle of Wight PO30 1RD ☎0906 550000 (premium) (07.00-22.00) Fares, road & rail hotline. ☎0800 961030 (freephone) comment line. National Bus info. GB Bus Timetable £12 per issue (3 per year) ☎01903 522456.

LEEP (Lothian & Edinburgh Environmental Partnership) ☎0131 468 8658. <www.leep.org.uk>. Car club advice.

Liquid Petroleum Gas Association <www.lpga.co.uk>.

Living Lightly Ltd, 14 Holly Terrace, York YO1 4DS. ☎01904 672489. <www.livinglightly.co.uk> (under construction) <mark@livinglightly.co.uk>. Bike trailers for luggage.

Local Exchange Trading (LETS), LETSLink UK, Basement Flat, 54 Campbell Rd, Portsmouth, Hampshire PO5 1RW ☎& Fax 023 9273 0639 <www.letslinkuk.org> <lets@letslinkuk.org> Barter trading.

Local Government Association (LGA), Local Government House, Smith Square, London SW1P 3HZ ☎020 7664 3000. <www.lga.org.uk>. Organise Don't Choke Britain month (June).

London Car Share <www.london-carshare.co.uk>.

London Cycling Campaign, 228 Gt Guildford Business Sq, 30 Gt Guildford St, London SE1 0HS ☎020 7928 7220 Fax 020 7928 2318. <www.lcc.org.uk> <office@lcc.org.uk>. Publish *London Cyclist*, an all abilities cycling directory and cycle maps.

London Recumbents, Rangers Yard, Dulwich Pk, London SE21 7BQ ☎020 8299 6636. <recumbents@aol.com>. Staff Yard, Battersea Park, London SW11 4NJ ☎020 7223 2533. Disability cycling & hire.

London Transport Museum, Wellington St, Covent Gdn, London WC2E 7BB ☎020 7379 6344 Fax 020 7565 7254. <www.ltmuseum.co.uk> <contact@ltmuseum.co.uk>.

London Transport ☎020 7222 1234 daily (24 hrs) Travel Check ☎020 7222 1200. Underground & bus.

London Underground, 55 Broadway, London SW1H 0BD.

LTS Rail, Central House, Clifftown Rd, Southend on Sea, Essex SS1 1AB ☎08457 444422 (local) tickets, updates ☎08457 678765 (local).

Mailing Preference Service, Freepost 22, London W1E 7EZ ☎0345 034599 (local). Anti-junk mail registration for five years.

MapInfo Ltd, Minton Place, Victoria St, Windsor, Berks SL4 1EG ☎01753 848200 Fax 01753 621140. <www.mapinfo.com> <uk@mapinfo.com>. Geographical information systems plot locational data e.g. postcodes to help understand markets & customers.

Members of Parliament (MPs), House of Commons, London SW1A 0AA ☎020 8219 4272. <www.parliament.uk> for email addresses.

Merseyrail Electrics, Rail House, Lord Nelson St, Liverpool L1 1JF ☎0151 709 8292 Fax 0151 702 2671. Trains.

Met Office ☎0336 401 932 (premium) Fax 0336 415 701. Weatherwatch forecast.

Midland Mainline, Midland House, Nelson St, Derby, E Mids DE1 2SA ☎08457 125678 (local) daily (08.00-20.00) Tickets. Special needs ☎0114 2537654. <www.midlandmainline.com>. Trains.

National Asthma Campaign, Providence House, Providence Pl, London N1 0NT, Bike for Breath, Freepost Lon 1120 London N1 0NC Helpline ☎08457 010203 (local) (M-F 09.00-19.00) Admin ☎020 7226 2260 Fax 020 7704 0740. <www.asthma.org.uk>.

National Cabline ☎0800 123444 (freephone). Taxis.

National CarShare, 8 Cressida Chase, Warfield, Bracknell, Berks RG42 3UD ☎01344 861600 Fax 0870 130 6773. <www.nationalcarshare.co.uk> <enquiries@nationalcarshare.co.uk>. Car share information service.

National Cycle Register ☎01244 539990. £5 per cycle to register against theft (2000).

National Driver Information & Control System <www.nadics.org.uk>. Scottish traffic control site.

National Express Ltd, Travel Sales Centre, Spencer House, Digbeth, Birmingham B5 6DQ ☎08705 808080 (national) (08.00 - 22.00), 5 days by post. <www.nationalexpress.co.uk> Cheques (£30+) to "National Express Ltd". Coaches to 1,200+ UK destinations.

National Federation of Shopmobility UK, 85 High St, Worcester WR1 2ET ☎ & Fax 01905 617761. <shopmob@dircon.co.uk>. Electric scooter loans.

National Parks (Association of) **(ANPA)**, Prof Ian Mercer, ANPA, Ponsford House, Moretonhampstead, Devon TQ13 8NL ☎01647 440245 Fax 01647 440187. <www.anpa.gov.uk> <IanMercer@anpa.gov.uk>.

National Society for Clean Air & Environmental Protection (NSCA), 136 North St, Brighton BN1 1RG ☎01273 326313 Fax 01273 735802. <www.greenchannel.com/nsca> <admin@nsca.org.uk>. Charity dealing with environmental pollution.

National Trust, Membership Dept, PO Box 39, Bromley, Kent BR1 3XL ☎020 8315 1111 Fax 020 8466 6824. <www.nationaltrust.org.uk> <enquiries@ntrust.org.uk>. Historic properties: see 'places to visit' list & call properties for travel details.

Natural Gas Vehicles Association, 11 Berkeley Street, Mayfair, London W1X 6BU. ☎020 7355 5086 Fax 020 7355 5099.

Northern Spirit, PO Box 208, Leeds LS1 2BU ☎08457 484950 (local) Fax 0113 245 2219. <www.northern-spirit.co.uk> <customerservices.ns@ems.rail.co.uk>. Trains.

Office of Fair Trading (OFT), Fleetbank House, 2-6 Salisbury Square, London EC4Y 8JX. ☎0345 224499 (local). <www.oft.gov.uk>. Consumer complaints.

Offroad Cycling, Coddington, Ledbury HR8 1JH ☎01531 633500 Fax 01531 636247. <colin-palmer@branchline.demon.co.uk>. Consultancy, training in reducing car use.

On Your Bike, EMAP Active, Apex House, Oundle Rd, Peterborough PE2 9NP ☎01733 898100. <www.onyourbike.com> <editorial@onyourbike.com>. Family magazine for new & born again cyclists, £2.75 (2000).

Open Road, The Danesmead Wing, 33 Fulford Cross, York YO10 4PB ☎01904 654654 Fax 01904 654684. Subscriptions/orders ☎01482 880399 <www.bycycle.com> <www.bikeculture.com> <peter@bcqedit.demon.co.uk>. Cycle consultants. Publish *Bike Culture Quarterly*, *Bycycle* & *Encycleopedia*.

Orbit Cycles, Unit 18 City Rd Trading Est, 295 City Rd, Sheffield S2 5HH ☎0114 275 6567. <www.orbit-cycles.co.uk>. German load carrying trailers. Cycle manufacturer.

Ordnance Survey (OS), Romsey Road, Maybush, Southampton SO16 4GU. ☎08456 050505 (local). <www.ordsvy.gov.uk>. Maps.

Organics Direct ☎020 7729 2828 <www.organicsdirect.co.uk>.

Outdoor Industries Association, Morritt House, 58 Station Approach, South Ruislip, Middx HA4 6SA ☎020 8842 1111 Fax 020 88420090. <info@go-outdoors.org.uk>. Trade body.

Parliamentary Advisory Council for Transport Safety (PACTS), St Thomas' Hospital, Lambeth Palace Rd, London SE1 7EH ☎020 7922 8112/3 Fax 020 7401 8740. <www.pacts.org.uk> <admin@pacts.org.uk>. Transport safety advisors.

Passenger Transport Networks, 49 Stonegate, York YO1 8AW ☎01904 611187. <jtyler@ptn.globalnet.co.uk>. Consultancy using Geographic Information Systems (e.g. postcode-based maps) to understand demand for travel by bus & train, & therefore plan & market services.

Pass-Plus Board, c/o Association of British Insurers, 51 Gresham St, London EC2V 7HQ ☎0115 901 2633 Fax 0115 901 2600. <www.passplus.org.uk>. Training scheme by the Driving Standards Agency for newly qualified drivers for safety & to save on insurance premiums.

Pedestrians Association (PA) the / Walk to School, 31-33 Bondway, London SW8 1SJ ☎020 7820 1010 Fax 020 7820 8208. <www.pedestrians.org.uk> <www.walktoschool.org.uk> <info@pedestrians.org.uk>. Charity to protect & promote the rights of walkers & its benefits. Publish *Walk* & *Walk to School* magazines. Free info with A5 SAE. Members £15 p.a. concs £10, family £20 (2000).

***Permaculture* Magazine**, The Sustainability Centre, East Meon, Hampshire GU32 1HR ☎01730 823311 Fax 01730 823322. <www.permaculture.co.uk> <info@permaculture.co.uk>. *Earth Repair Catalogue*, books & Ecoflow fuel conditioner.

Primal Seeds ☎0161 224 4846. <www.primalseeds.org> <mail@primalseeds.org>. Organics.

RAC Motoring Services, Great Pk Rd, Bradley Stoke, Bristol BS32 4QN ☎0800 029 029 (freephone). <www.rac.co.uk>. Free internet route planner.

Rail Enquiries UK ☎08457 484950 (local 24 hrs) <www.railtrack.co.uk/travel>. Docklands Light Railway ☎020 7363 9700 <www.dlr.co.uk> <cservice@dlr.co.uk>. Glasgow Underground ☎0141 332 7133 daily (07.00-22.30) Tyne & Wear Metro ☎0191 232 5325. Manchester Metrolink ☎0161 205 200. Merseyside ☎0151 236 7676. West Yorks Metroline ☎0113 245 7676. National Rail Timetable from stations, W.H. Smith or Teamwork Direct, 5 Chessingham Pk, Dunnington, York YO19 5YA ☎01904 481140.

Rail Europe Direct, 179 Piccadilly, London W1U 0BA ☎0990 848848 (national). <www.raileurope.com> <info@raileurope.com>. Train tickets.

Rail Planner, Tranley House, Tranley Mews, 144 Fleet Rd, London NW3 2QW ☎020 7267 7055. Fax 020 7267 2745. <www.railplanner.com> <enquiries@railplanner.co.uk>. Route planning software £45 p.a, TubePlanner £9.99 (2000).

Railtrack PLC, Euston Sq, London NW1 2EE. Helpline ☎08457 114141 (local, 24 hours). Track works information.

Rail Users Consultative Committees (Eastern England) Crescent House, 46 Priestgate, Peterborough PE1 1LF ☎01733 312188. (London) Clements House, 14-18 Gresham St, London EC2V 7PR ☎020 7505 9000. (Midlands) 6th Floor, McLaren Building, 35 Dale End, Birmingham B4 7LN ☎0121 212 2133. (N E England) Hilary House, 16 St Saviour's Pl, York YO1 7PJ ☎01904 625615. (NW England) Boulton House, 17-21 Chorlton St, Manchester M1 3HY ☎0161 228 6247. (Scotland) Rm 514, Corunna House, 29 Cadogan St, Glasgow G2 7AB ☎0141 221 7760. (Southern England) 4th Fl, 35 Old Queen St, London SW1H 9JA ☎020 7222 0391. (Wales) St Davids House, Wood St, Cardiff CF1 1ES ☎01222 227247. (Western England) Tower House, Fairfax St, Bristol BS1 3BN ☎0117 926 5703. Watchdog for rail users.

Raleigh ☎0115 942 0202. Cycles.

Ramblers Association, 1-5 Wandsworth Rd, London SW8 2XX ☎020 7339 8500 Fax 020 7339 8501. <www.ramblers.org.uk>. Walking charity.
Reclaim the Streets, PO Box 9656, London N4 4JY ☎020 7281 4621. <www.reclaimthestreet.net> <rts@gn.apc.org>. Direct action against car culture.
Retail Motor Industry Federation National Conciliation Service, 9 North St, Rugby CV21 2AB ☎01788 576465. Complaints about used cars, repairs & servicing in England, Wales & Northern Ireland.
RoadPeace, PO Box 2579, London NW10 3PW. Helpline ☎020 8964 1021 daily (09.00-21.00) Office ☎020 8838 5102 Fax 020 8838 5103. <www.roadpeace.org.uk> <info@roadpeace.org.uk>. Charity offering emotional & practical support to victims & campaigning for safer roads. Publish newsletters & *Safety First* magazine. Membership by donation.
Royal Mail Post code check ☎0345 111 222 (local). Services (e.g. Post bus info) ☎0345 740740 (local). Textphone ☎0845 600 0606. Parcelforce ☎0800 224466 (freephone). Business ☎0345 950 950 (local). <www.royalmail.co.uk>.
Royal Society for the Prevention of Accidents (RoSPA), Edgbaston Park, 353 Bristol Rd, Edgbaston, Birmingham B5 7ST ☎0121 248 2000 Fax 0121 248 2001. <www.rospa.co.uk> <help@rospa.co.uk>. Safety charity & training: Advanced Driving Test & Diplomas in Advanced Driving & Riding Motorcycles.
Sainsbury's Supermarkets Ltd, Stamford House, Stamford St, London SE1 9LL. Careline ☎0800 636232 (freephone). Home shopping ☎0845 3012020 (local). <www.sainsburys.co.uk>. £5 delivery charge (2000).
ScotRail Railways, Caledonian Chambers, 87 Union St, Glasgow G1 3TA ☎08457 550033 (local) daily (08.00-22.00) tickets. Customer services ☎0141 335 4240. <www.scotrail.co.uk>.
Scottish City Link, Customer Services, Buchanan Bus Station, Killermont Street, Glasgow G2 3NP. ☎08705 505050 (1 day by post) Fax 0141 332 4488. <www.citylink.co.uk> <info@citylink.co.uk. Coaches to 200 Scottish destinations.
Scottish Motor Trade Association (SMTA), 3 Palmerston Pl, Edinburgh EH12 5AF ☎0131 225 3643 Fax 0131 220 0446. <www.smta.co.uk> <billdunn211@netscapeonline.co.uk>. Complaints form about used cars, repairs & servicing in Scotland.
Scottish Youth Hostel Association (SYHA), 7 Glebe Cres, Stirling FK8 2JA ☎01786 891400 Fax 01786 891333. <www.syha.org.uk>.

Semlyen Anna, c/o Green Books, Foxhole, Dartington, Totnes, Devon TQ9 6EB. <www.cuttingyourcaruse.co.uk> <info@cuttingyourcaruse.co.uk>. Cutting Your Car Use Consultant, Researcher with **Passenger Transport Networks** & author. Personal transport & car costs advice, site-specfic travel guides, Green Transport Planning & Ecoflow fuel conditioners.

Silverlink Train Services Ltd, Melton House, 65-67 Clarendon Rd, Watford WD1 1DP ☎08705 125240 daily (08.00-20.00). <www.silverlink.co.uk>. Tickets.

Sinclair Research, 7 York Central, 70 York Way, London N1 9AG. ☎020 7837 6150. <www.sinclair-research.co.uk>. ZETA III power assistance for cycles.

Slower Speeds Initiative (SSI), PO Box 746, Norwich NR2 3LJ ☎ & Fax 01603 504563. <www.slower-speeds.org.uk>. Road safety charity & campaign coalition.

Smart Moves Ltd, Technocentre, Puma Way, Coventry CV1 2TT ☎024 7623 6292 Fax 024 7623 6291. <www.smartmoves.co.uk> <info@smartmoves.co.uk>. Support & consultancy on personal transport & pay-as-you-drive car clubs. Publish *The Car Club Kit Book* on car sharing £7.50 (2000).

Society of Motor Manufacturers & Traders (SMMT), Customer Relations, Forbes House, Halkin St, London SW1X 7DS ☎020 7235 7000 Fax 020 7235 7112. <www.smmt.co.uk> <communications@smmt.co.uk>. Complaints about cars under manufacturer's warranty.

Soil Association, Bristol House, 40-56 Victoria St, Bristol BS1 6BY ☎0117 929 0661 Fax 0117 925 2504. <www.soilassociation.org> <info@soilassociation.org>. Charity promoting organic food and farming. Local food links pack.

South West Trains, Friars Bridge Court, 41-45 Blackfriars Rd, London SE1 8NZ ☎0845 6000650 daily (08.00-20.00) <www.swtrains.co.uk>. Tickets.

Stagecoach Oxford ☎01865 772250. <www.stagecoach-oxford.co.uk> <info@stagecoach-oxford.co.uk>. Oxford-London coaches every 12 mins.

Stagecoach Supertram Sheffield, Nunnery Depot, Woodbourne Road, Sheffield S9 3LS ☎0114 272 8282. Trams.

Stanfords, 12-14 Long Acre, Covent Garden, London WC2E 9LP ☎020 7836 1321. <www.stanfords.co.uk> <sales@stanfords.co.uk>. Maps.

Stationery Office (The), St Crispins, Duke Street, Norwich NR3 1PD ☎0870 600 5522 Fax 0870 600 5533. <www.itsofficial.net> The *Highway Code* & *Planning Policy Guidance Notes*.

Stirling Surveys ☎01786 479866 Fax 01786 472914.
<www.stirlingsurveys.co.uk> <info@stirlingsurveys.co.uk>. Footprint &
cycle maps.
Stop Fuming, 110 St Martin's Lane, Covent Gdn, London WC2N 4DY
☎020 7841 5469 Fax 020 7841 5777. <lgoddard@cbarker.co.uk>.
Motor industry campaign for regular car servicing.
Sustrans, PO Box 21, Bristol BS99 2HA. Info/sales ☎0117 929 0888
Fax 0117 915 0124. <www.sustrans.org.uk> <info@sustrans.org.uk>.
Practical charity for walking & cycling. National Cycle Network, Safe
Routes to Schools, free catalogue of maps & technical publications.
Tandem Club, 25 Hendred Way, Abingdon, Oxfordshire OX14 2AN
☎01235 525161 Fax 01235 445706. <www.tandem-club.org.uk>
<p.j.hallowell@rl.ac.uk>. Magazine & handbook, club rides.
Taylor Cycles, 375 Birchfield Rd, Reddich, Worcs B97 4NE ☎01527
545262. <www.cycling.co.uk>. Trailer cycles, road cycles for the disabled.
Technicolour Tyre Co, PO Box 373, Brookwood, Woking, Surrey GU24
0BA ☎01483 797675 Fax 01483 797681. Mail order safety reflectives.
Telephone Preference Service, 5th Fl, Haymarket House, 1 Oxendon
St, London SW1Y 4EE ☎0845 070 0707 (local). <www.dma.org.uk>
<tps@dma.org.uk>. Blocking unwanted calls.
TCA (Telework, Telecottage & Telecentre Association), Freepost
CV2312, Wren, Kenilworth, Warwicks CV8 2RR ☎0800 616008 (freephone)
Fax 024 7669 6538. <www.tca.org.uk> <teleworker@compuserve.com>.
Not-for-profit organisation for flexible working. *Teleworker* magazine (6
p.a.), handbook. Individual £34.50 p.a. (2000).
Tesco Stores Ltd, PO Box 73, Baird Avenue, Dundee DD1 9NF ☎0800
505555 (freephone) Fax 0138 281 9956. <www.tesco.co.uk>
<customer.services@tesco.net>. Tesco Direct home shopping (10.00-
22.00 Mon-Sat). £5 per order.
TGA Electric Leisure, Factory Lane West, Halstead, Essex CO9 1EX
☎01787 478430 Fax 01787 478432. Electric bikes.
Thameslink Rail, Friars Bridge Court, 41-45 Blackfriars Rd, London SE1
8NZ ☎020 7620 6333. <www.thameslink.co.uk>.
Thames Trains, Venture House, 37 Blagrave St, Reading RG1 1PZ
☎08457 300700 for tickets. <www.thamestrains.co.uk>.
Thomas Cook Publishing, Dept TPO/MAP PO Box 227,
Peterborough PE3 8BQ. <www.thomascook.co.uk>
<talktous@thomascook.com>. Rail maps & European timetable.

Thomson Directories Ltd, Thomson House, 296 Farnborough Rd, Farnborough, Hampshire GU14 7NU ☎0800 9750832 (freephone). <www.thomweb.co.uk> <comments@thomweb.co.uk>.
Traidcraft Plc, Kingsway, Team Valley Trading Estate, Gateshead, Tyne & Wear NE11 0NE ☎0191 491 1001. <www.traidcraft.co.uk>. Newsletter <ruthb@traidcraft.co.uk>. Home delivery of fairly traded goods.
Train Line <www.thetrainline.com>. Enquiries/bookings.
Tramtrack Croydon, Tramlink Info Centre, Unit 5, Suffolk House, George St, Croydon ☎020 8681 8300 Fax 020 8760 5655. <www.tramlink.uk.com>.
Transpenninexpress Rail, Northern Spirit Ltd, Customer Services, PO Box 208, Leeds LS1 2BU ☎0870 6023322 (national). Wheelchair booking ☎0845 600 8008 (local) give 24 hrs notice.
Transport 2000, The Impact Centre, 12-18 Hoxton St, London N1 6NG ☎020 7613 0743 Fax 020 7613 5280. Traffic reduction campaign working with government, corporations & with a network of local groups, *Transport Retort* magazine; good practice guides for journeys to work, school, hospitals and (forthcoming) leisure sites; also campaigning guides. Supporters £20 p.a. (2000); ask about concs.
TravelWise Association (National) **(NTWA)**, Brian Smith, Director of Environment & Transport, Cambridgeshire County Council, Castle Court, Shire Hall, Castle Hall, Cambridge CB3 0AP ☎01223 717790 Fax 01223 718458 <www.travelwise.org.uk> <brian.smith@envtran.camcnty.gov.uk>. Campaign to change attitudes to car use.
Trek, Unit B, Maidstone Road, Kingston, Milton Keynes MK10 0BE ☎01908 282626 Fax 01908 285006. Cycle trailers.
Tripscope ☎08457 585641 <tripscope@cableinet.co.uk>. Elderly & disabled travel helpline.
Tug Co, The Barn, Church Lane, Clyst St Mary, Exeter EX5 1AB. ☎ & Fax 01392 877872. UK cycle trailers.
Two Plus Two see **Bike Hod**.
UK Public Transport Information <www.pti.org.uk>.
UK Trailer Co, The Camel Trail, Wadebridge, Cornwall PL27 7AL ☎07000 287539 Fax 01208 814407. <burley@uktrailerco.freeserve.co.uk>. Burley cycle trailers for young children & loads.
Valley Cycles, Unit 2 Nene Court, The Embankment, Wellingborough NN8 1LD ☎01933 271030 Fax 01933 271060. <www.valleycycles.co.uk> <sales@valleycycles.co.uk>. Bike Friday & Family Tandem folding cycles.

Valley Lines, Brunel House, 2 Fitzalan Rd, Cardiff CF2 1SA ☎02920 449944 (local) update. Trains.

Vehicle Certification Agency (The), The Eastgate Office Centre, Eastgate Rd, Bristol BS5 6XX ☎0117 952 4126. <www.detr.gov.uk/gtp/index.htm>. New car fuel consumption figures

Vehicle Inspectorate ☎020 8665 0885. Smoky vehicle hotline. Trade association for MOT testing.

Virgin Trains, 85 Smallbrook Queensway, Birmingham B5 4HA. ☎08457 222 333 (local) daily (08.00-22.00) tickets. Fax 0121 654 7500. <www.thetrainline.com> <www.virgintrains.co.uk>. Assistance ☎08457 443366 (local) with 24 hrs notice.

Wales & West, Brunel House, 2 Fitzalan Rd, Cardiff CF4 0SU ☎08457 125625 tickets. <www.walesandwest.co.uk>. Trains.

West Anglia Great Northern Railway (WAGN), Station Rd, Cambridge CB1 2JW ☎0800 566566 (freephone) tickets, updates ☎08457 445522 (local).

West Coast Railway Co, Warton Rd, Carnforth, Lancs LA5 9HX ☎01463 239026. Summer steam trains.

Wheel Alternatives Ltd / E Trans, 40 Lincoln St, York YO26 4YR ☎01904 338338. <wheel.alternatives@workbike.org> <www.workbike.org>. Couriers, loads & wedding cycle transport.

Wheels for All, 1 Enterprise Park, Agecroft Road, Pendlebury, Manchester M27 8WA ☎0161 745 9099. <www.cycling.org.uk>. Will be publishing a directory for all abilities.

Yellow Pages, Queens Walk, Reading RG1 7PT ☎0800 671444 (freephone). <www.yell.co.uk>.

York Cycleworks, 14-16 Lawrence St, York YO10 3WP ☎01904 626664. <www.york-cycleworks.co.uk>. Mail order trailers.

Youth Hostels Association (YHA), England & Wales, 8 St Stephen's Hill, St Albans AL1 2DY ☎01727 845047 Fax 01727 844126. <www.yha.org.uk>

Zero Emissions ☎020 7723 2409 <www.workbike.org/zero> <zero@workbike.org>. Cycle couriers & deliveries.

Distance, speed and fuel conversion tables

1 yard = 91.4 centimetres 1 metre = 3.279 feet
1 kilometre = 1090 yards 1 kilometre = 0.6213 miles
1 mile = 1620 yards   1 mile = 1.6093 kilometres

Speeds	mph	10	20	30	40	50	60	70
	kmph	16	32	48	64	80	96	112

Fuel Consumption (1 gallon = 4.546 litres)

mpg	10	15	20	25	30	35	40	45	50
miles/10 litres	22	33	44	55	66	77	88	99	110

AUTOMOBILES:
THE MYTH
THE REALITY
SINGER

Feedback and updates

The author mixing
modes of travel

Feedback is *very welcome*. It will improve future editions and help to keep my work up to date. Traffic reduction is a rapidly innovating subject area!

I collect advice and research on personal mobility and on being voluntarily car-free. Special interests are green travel training, personal transport economics, mixed modal guides and Green Transport Plans.

Please tell me about organisations you think should be in the directory, new addresses or about services or products which help to cut car use. Your experiences are also welcome.

info@cuttingyourcaruse.co.uk
Anna Semlyen, c/o Green Books,
Foxhole, Dartington, Totnes, Devon TQ9 6EB, UK.

Top ideas will be posted at <www.cuttingyourcaruse.co.uk>.
Please *show this book* to people whose work involves reducing traffic and *ask your library to buy a copy*. Please also *link to my web site*.

JOURNEY DIARY

• Complete each leg of the journey separately, as some may involve many modes.

Day	Time	From	To	Miles	Aim(s)	Mode	Cost £	Trip Time
1								
2								
3								
4								
5								
6								
7								
8								
9								
10								

JOURNEY DIARY

Plan alternatives to the car for any regular or repeat trips. Pages 48-75 on modes will help. Get a map, directions or timetables if needed. See pages 28-31.

Alternatives to car – phone/write/localise/delivery/walk/ cycle/public transport/lift share/taxi/pool/car hire/moped ...	Frequency time, stops	Cost £	Trip time
1			
2			
3			
4			
5			
6			
7			
8			
9			
10			

JOURNEY DIARY (continued)

Day	Time	From	To	Miles	Aim(s)	Mode	Cost £	Trip Time
1								
2								
3								
4								
5								
6								
7								
8								
9								
10								

JOURNEY DIARY (continued)

Alternatives to car – phone/write/localise/delivery/walk/ cycle/public transport/lift share/taxi/pool/car hire/moped ...	Frequency time, stops	Cost £	Trip time
1			
2			
3			
4			
5			
6			
7			
8			
9			
10			

Sustrans National Cycle Network

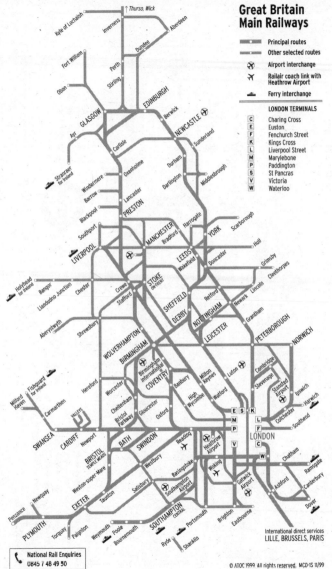

Great Britain
Main Railways

- Principal routes
- Other selected routes
- ✈ Airport interchange
- ✈ Railair coach link with Heathrow Airport
- ⛴ Ferry interchange

LONDON TERMINALS

C	Charing Cross
E	Euston
F	Fenchurch Street
K	Kings Cross
L	Liverpool Street
M	Marylebone
P	Paddington
S	St Pancras
V	Victoria
W	Waterloo

International direct services
LILLE, BRUSSELS, PARIS

National Rail Enquiries
0845 / 48 49 50

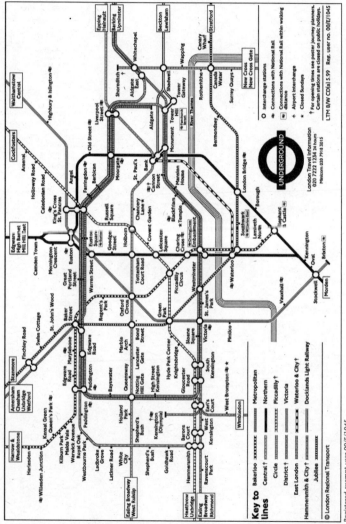

Key to lines

Bakerloo
Central †
Circle
District †
East London
Hammersmith & City †
Jubilee
Metropolitan
Northern
Piccadilly †
Victoria
Waterloo & City †
Docklands Light Railway

○ Interchange stations
⊖ Connections with National Rail
⊕ Connections with National Rail within walking distance
✈ Airport interchange
★ Closed Sundays
† For opening times see poster journey planners. Certain stations are closed on public holidays.

London Travel Information
020 7222 1234 24 hours
Minicom 020 7918 3015

LTM B/W CD(a) 5.99   Reg. user no. 00/E/1045

© London Regional Transport

Registered exempt user 00/E/1045

Index

air quality 16, **43-45**
asthma 15, **43-45**
blading **72-73**
buses 15, 26, 29, **30-31**, 43,
  61-62, 65, 78, 84, 112
business journeys 23, 62, 65,
  69, 77, **79-80**, 94
cars & car use
  allowances 79, **94**
  clubs 14, 17, 34, **68-70**,
    101
  costs 7, **13-14**, 35, 45, 67,
    68, **94-106**
  giving up your car **101-104**
  hire 53, **67**, 69, 114
  pools 34, **69**, 79
  purchase criteria **106**
  scrapping or selling **103**x
  sharing **50-52**, **68**, 78, 82,
    84
  small cars 14, 67, **105**
  see also lift sharing,
    park & ride, ride sharing,
    speed
children 9, 15, 18, 33, 35, 43,
  54, 56, 64, 66, **80-84**, 110,
  115, **116-117**
coaches 30, **66**, 109
commuter journeys 9, 38, 52,
  77-78, 91
congestion 8, 18, **38-39**, 44,
  83, 111

cycling 14, 15, 18, 25, 26,
  27, 28, 33, 34, 36, 37, 43,
  53-60, 77, 78, 79, 80, 82,
  83, 85, 88, 109, 111-115,
  156
  costs **91-92**
  electric-powered *see*
    electric bicycles
  folding *see* folding bicycles
  purchase criteria **57-58**
  security 54, **59**
  see also trailers, tricycles
directions 23, **28-29**, 39, 49,
  52, 78
disabilities 49, 62, 64
  cycling 54
distance 21, 23, 25, 28, **150**
electric bicycles **60**
energy efficiency **40-43**, 46,
  60, 105, 106, 150
environment 7, **16-17**, **43-46**,
  57, 106
flexi-time 18, 23, 25, 38, **77**
folding bicycles 14, 18, **54**,
  62, 65, 78, 79, 112, 114
fuel **40-43**, **45-46**, 106, 150
health 7, 14, **15**, 26, **33-37**,
  44, 49, 53, 54, 72
hitch-hiking 52, 58, **73**
horses **74**
journey
  diary **152-155**

planning 14, **74-75**, 101
leisure journeys **88-89**
lift sharing 18, **50-2**, 75, 113
loads 23, 28, 56, 105, **113-115**
localising 16, 18, **25**, 38, 88, 117-118
maps **28-9**, 49, 55, 56, 73, 82, 101, **156-158**
mixed transport modes 91, **93-94**
mopeds **70-72**
motorbikes **70-72**
noise **46-47**, 70, 88, 89
park & ride 39, **61-62**
parking **39 40**
pollution **40-47**, 88-89
ponies **74**
quiet **46-47**, 54, 80
ride sharing **50-52**
roller blading **72-73**
roller skating **72-73**
safety 18, **33-37**, 70, 72, 80, **81**, 82-83, 106, **109-110**
saving money **13-14**, 25, **40-43**, 49, 50, 52, 66, 68, 70, **91-107**
school journeys 15, 38, **80-84**
scooters **70-72**
shopping 18, 21, **24**, **85-87**, 101, **113-115**
skate boarding **73**
socialising 18, 25, 49, **87-88**

speed 18, 26, 27, **33-34**, 35-37, 41, 46, 82, 105, **150**
staying still **23-24**
targets **117-118**, 119
taxis 18, 50, **52-53**, 65, 69, 77, 78, 85, 86, 101, 103, 111, 114
telecommuting **23**
teleconferencing **23**, 78
timetables **30-31**
trailers
    car 105
    cycle 79, 101, **115**, 117
trains **63-66**, 110, **156-8**
tricycles 54, 70, 115
underground **63-66**, 158
walking 15, 25, 28, 36, 38, 43, **49**, 78, 80-3, 85, 87, 88, 93, 109-110, 111-114, 116